W9-CLC-632

100 DIVES OF A LIFETIME

100 DIVES OF A LIFETIME

The World's Ultimate Underwater Destinations

CARRIE MILLER

WASHINGTON, D.C.

Since 1888, the National Geographic Society has funded more than 13,000 research, exploration, and preservation projects around the world. National Geographic Partners distributes a portion of the funds it receives from your purchase to National Geographic Society to support programs including the conservation of animals and their habitats.

National Geographic Partners
1145 17th Street NW
Washington, DC 20036-4688 USA

Get closer to National Geographic explorers and photographers, and connect with our global community. Join us today at nationalgeographic.com/join

For information about special discounts for bulk purchases, please contact
National Geographic Books Special Sales: specialsales@natgeo.com

For rights or permissions inquiries, please contact
National Geographic Books Subsidiary Rights: bookrights@natgeo.com

Library of Congress Cataloging-in-Publication Data
Names: Miller, Carrie, (Writer), author.
Title: 100 dives of a lifetime : the world's utlimate underwater destinations / Carrie Miller ; foreword by Brian Skerry.
Other titles: One hundred dives of a lifetime | Hundred dives of a lifetime
Description: Washington, DC : National Geographic, [2019] | Includes index.
Identifiers: LCCN 2018027276 | ISBN 9781426220074
Subjects: LCSH: Scuba diving--Handbooks, manuals, etc. | Underwater exploration.
Classification: LCC GV838.672 .M45 2019 | DDC 797.2--dc23
LC record available at https:// lccn.loc.gov_2018027276

Printed in China

18/RRDS/1

Scuba diving is a potentially dangerous activity that can expose a diver to significant risks, including serious injury or even death. Important factors for safe diving include, but are not limited to, optimal health and physical fitness, specialized training, diving certification, specialized equipment, experience, and safe diving practices. The content of this book is for information purposes only, and is not intended as a replacement for the above safe-diving factors, nor as encouragement for anyone to scuba dive in the absence of the above safe-diving factors and any additional factors that may be specific to the individual diver, the dive location, or the dive environment. Before engaging in any scuba diving activities we recommend that you consult with a certified scuba diving instructor to help you determine the specific steps you should take to have a safe diving experience.

Any products, services, processes, or website content referenced in this book is provided for information purposes only and does not constitute an endorsement or recommendation by National Geographic or its affiliates.

By electing to use any information contained in this book, you understand and agree that neither National Geographic nor its affiliates make any warranty, express or implied, including the warranties of merchantability and fitness for a particular purpose, or assume any legal liability or responsibility for any information, products, services, processes, or website content referenced in this book, and are not liable for any loss; damage; or injury, including death, that may result from your election to engage in any scuba diving activities and/or use any products or services or information mentioned herein. ❍

CONTENTS

Pages 2–3: An explosion of marine life surrounds this coral reef 66 feet (20 m) below the waters of Mohéli Marine Park in the Comoros.

Left: Emperor penguins swim beneath the ice floes of Antarctica.

FOREWORD

I became a certified scuba diver when I was 16 years old. Although I didn't know it then, my life would never be the same. Receiving my "C-card" was like being given a key to a hidden, parallel universe filled with wonderful creatures and remarkable alien landscapes.

I remember my very first ocean dive in the waters off Jamestown, Rhode Island. The chilly water trickled in around the neck of my wet suit and down my back, but I was far too mesmerized by all that lay before me to really notice. With my dive instructor leading the way, I descended the rocky wall, pausing to study orange and white *Metridium* anemones while cunner fish darted around a few feet beyond. I peered into crevices and found crabs and lobsters returning my gaze. The dive lasted perhaps 30 minutes, but I felt as though I had traveled to another world. I remember seeing people on the beach later that day and feeling as though I knew something they did not. Something had been revealed to me that no one else knew. I could barely contain my excitement. In my heart, I knew that something had changed.

Forty years later I realize that my instincts were correct. One brief scuba dive in cool, green water led to a lifetime journey of exploration and discovery that has opened my eyes and enriched my soul in ways I could have never imagined.

As terrestrial dwellers, we understandably often see our world from a land-centric perspective. But Earth is a water world, where nearly three-quarters of its surface shimmers and pulses with tides. Perhaps even more noteworthy is that 98 percent of our planet's biosphere, the place where life can exist, is water. Given that exploration is in our DNA, we would miss so much if our treks were relegated to only dry land. For those who dive, the water's edge is never the end of the trail; it is just the beginning. Beneath the waves awaits a realm quite unlike any other on Earth, where on any given day absolute magic can be found.

Throughout my career, I've never ceased to be amazed at the extraordinary experiences I continue to have underwater. On land, getting close to wildlife seems especially difficult. I can rarely get near a squirrel or bird in my backyard. Underwater, however,

Diver Vincent Canabal makes sure a tiger shark at the Bahamas' Tiger Beach keeps a respectful distance.

animals often seem more tolerant, even curious at times. In locations worldwide, we can swim with dolphins, mantas, sea lions, and sharks. Cruise alongside a whale shark in Mexico or free-dive with dolphins in the Bahamas, and you'll be hooked forever. Before you know it, you'll be pirouetting with stingrays in the Caymans, surrounded by scalloped hammerheads in the Cocos, and blowing bubbles with belugas in Russia. Speaking from experience, I can say this stuff is addictive!

But the great thing about diving is that there is something for everyone and an endless number of possible experiences. Big animals and adrenaline-pumping dives are remarkable, but relaxing swims over stunning habitats bring a completely different type of joy. On a five-week expedition to the Southern Line Islands, I found myself in the water day and night. But my favorite time was dusk on the reef. Light levels dim, creating a powder blue seascape as schools of parrotfish sweep over the primal reef in search of a safe place to rest overnight, while nocturnal creatures emerge from their dens. The experience was peaceful and serene, and I found myself happily awaiting the soothing feelings it produced each day.

The undersea world offers endless potential to stimulate our sense of adventure, and great dives can take many forms. Visiting historic shipwrecks underwater is like traveling back in time. Descending to wrecks in locations such as Chuuk Lagoon or Bikini Atoll, you'll be transported back to the turbulent days when they sank long ago, while marveling at how the sea has transformed them into gardens of life.

In the pages ahead, you will learn of the premier dives in our water world. From polar to temperate to tropical, and from wildlife to wrecks to cenotes, these 100 dives offer the most exhilarating diving experiences the planet has to offer. So dive into each chapter, then pack your mask and fins and take to the sea! ❍

"For those who dive, the water's edge is never the end of the trail; it is just the beginning. Beneath the waves awaits a realm quite unlike any other on Earth, where on any given day absolute magic can be found."

—BRIAN SKERRY

A cluster of feather duster worms
at Lighthouse Reef in Belize

INTRODUCTION

Oceans have historically been something that has separated us. *An ocean apart. Across the sundering seas. The oceans between us.* Now, more than ever before, we understand the sea connects us. It is our shared life support system, a borderless expanse linking one continent with the next. It is our collective history, witness to fledgling seafaring attempts that advanced to legendary voyages, ghastly battles, and staggering discoveries.

Yet, although the sea knows us, we hardly know her. The ocean covers 70 percent of our planet's surface, but more than 80 percent is still unexplored. We don't even *know* what we don't know. For explorers, that's a heady tonic. It's why we dive.

Divers know the pang of disappointment reaching the half-tank mark when there's still so much more to discover. Divers understand the rich wonder of seeing a different world—full of color and life—play out in front of their eyes, a world few people get to observe. This shared experience creates a community of fellow adventurers.

But if there is one commonality among scuba divers, it is their diversity: From macrophotographers to ice enthusiasts, cave divers to reef lovers, divers' passions are as varied as the ocean itself. This made compiling a list of the world's 100 best dive destinations a mighty challenge. We spoke with National Geographic's underwater photographers and Explorers, passionate divers all. We also canvassed legions of divers from around the world on their favorite places to dive, locations that offered unique experiences, squadrons of marine life, discoveries great and ghost-shrimp small.

The result was a list that is inspirational, exciting, and will no doubt be hotly contested. You'll find perennial favorites like Australia's Great Barrier Reef (page 204) and Belize's Great Blue Hole (page 384). We've also included some adrenaline pumpers (like the Aliwal Shoal sardine run, page 300), stuff of legends (like the fabled Bikini Atoll, page 368), eerie wrecks, and one-of-a-kind experiences. (A no-vis hunt for megalodon teeth, anyone? See page 268.) In between are plenty of enjoyable, fish-filled destinations to add to your bucket list, from New Zealand's Poor Knights Islands (page 124) to the Galápagos (page 270).

A school of old wives
swarms a sunken jetty post.

The book is separated into three sections: (1) Beginner Dives, sites worth seeing for any diver, but selected because they are ideal for those beginning their underwater explorations, divers who don't get to dive as often as they would like, or who simply prefer an easy cruise; (2) Intermediate Dives, which kick things up a notch with dives that truly showcase the world, from the Red Sea, to remote atolls, to cold-water wrecks, and are for divers with more experience who are comfortable with their skills and are ready to be inspired; and (3) Advanced and All-Level Dives, extreme and extremely rewarding for any diver seeking these adventures, plus a few places that offer up a little something for everyone, from cenotes to the emerald Andaman Sea.

We've also included tips on how we can better protect the underwater wonderland that we love so much. Not only is the ocean our playground, but it is also the beating heart of our planet. It supports life, regulates temperature, drives weather patterns. As divers, we should be leading the charge to protect the ocean. And the more we see, the more we understand what we don't know.

"For explorers that's the ultimate," says Brian Skerry. "On any given day you'll never know what you'll see, and for someone with a sense of curiosity, there's always something new to explore and discover. We haven't even scratched the surface."

We hope this book will motivate old hats to dust off their logbooks, as well as inspire legions of new ocean explorers. ○

Divers prepare for a night swim at Garden Eel Cove with manta rays in Kona, Hawaii.

ARCTIC
NORTH AMERICA
SOUTH AMERICA
PACIFIC OCEAN
ATLANTIC OCEAN
Thingvellir
Scapa Flow
Barkley Sound
Mission Silo
Bell Island
Émergence du Ressel
Monterey Bay
Santa Catalina Island
Azores
Discovery Bay
Cooper River
Devil's Den
Tiger Beach
Ray of Hope Wreck and the Shark Arena
El Bajón del Rio
First Cathedrals
Kona
Gordo Banks
Yucatán Cenotes
Afuera
El Boiler
Great Blue Hole
RMS *Rhone*
Charlie Brown Wreck
Salt River Canyon
Mary's Place, Roatan
South Water Caye
Critter Corner
Sisters Rocks
Islas Santa Catalina
Something Special
Cocos Island
Malpelo
Kiritimati
Darwin Island
Los Jardines de la Reina
Fernando de Noronha
Bloody Bay Wall
Ascension Island
Lake Titicaca
Tuamotu Islands
Easter Island
South Georgia
Antarctica

DIVE DESTINATIONS

PART ONE

BEGINNER DIVES

A free diver swims down
the entrance of The Passage,
a shallow cave in North Raja Ampat.

NEW CALEDONIA

CATHEDRAL

A sheer drop-off with cruising pelagics

AVERAGE WATER TEMP: 75°F (24°C) **AVERAGE VISIBILITY:** 66 feet (20 m)
AVERAGE DEPTH: 80 feet (24 m) **TYPE OF DIVE:** Open water, night, reef, and lagoon

Nature has been kind to New Caledonia. This French territory, located 750 miles (1,200 km) northeast of Brisbane, Australia, has an immense barrier reef—about 990 miles (1,600 km) wrapped around one of the world's largest lagoons (approximately 9,260 square miles/23,983 sq km).

As expected, this circle of life contains an exceptionally diverse ecosystem, with 350 species of coral (vase sponge, brain, sea fans, table) and 1,600 species of identified fish (including triggerfish, clownfish, and Napoleon wrasse), not to mention dugongs and nesting green sea turtles. In 2008, UNESCO declared these special lagoons a World Heritage site.

For divers, New Caledonia offers an array of diverse explorations including shallow lagoon dives, night dives, reef dives, open sea and underwater drift dives, like Cathedral (also known as The Cathedral or Cathedrale), one of New Caledonia's best.

Located near Hienghène, Cathedral is a deep rift with a series of winding ravines, caves, tunnels, and caverns. Bursts of colorful reef fish swarm over anemones, sea fans, and a healthy growth of hard and soft coral. Large pelagics like tuna, humpback parrotfish, groupers, and barracuda cruise the scene along the vertical wall, overlooking a 180-foot (55 m) drop.

Sheltered by the Doïman Reef, the Cathedral's abundance of intact corals makes for one of the most breathtaking dives you'll find in New Caledonia. ❍

What You'll See: Manta Rays • Eagle Rays • Tuna • Green Turtles • Groupers • Dugongs • Seahorses • Nautilus • Anemones • Brain Coral • Sea Fans • Whitetip Reef Sharks • Nudibranchs • Vase Sponge Coral • Parrotfish • Triggerfish • Spanish Mackerels

An aerial view of New Caledonia's
Grande Terre coral reef

ICELAND

THINGVELLIR

Touch the walls of two continental shelves.

AVERAGE WATER TEMP: 37°F (2.8°C) **AVERAGE VISIBILITY:** 330 feet (101 m)
AVERAGE DEPTH: 33 to 60 feet (10 to 18 m) **TYPE OF DIVE:** Cold water

This is the place where worlds collide. Literally. The North American and Eurasian tectonic plates run side by side in Thingvellir, and divers willing to brave cold fingers can touch geologic history, placing a hand on each continent.

Diving the Silfra fissure in Thingvellir National Park (just 40 minutes outside of Reykjavík) isn't difficult—just cold. (A dry suit and dry suit experience is required.) After jumping in via a platform, the first thing that strikes divers is the unsurpassed clarity of the water. The visibility of this glacial spring extends hundreds of feet, providing an uninterrupted view of the channel walls, which seem to emanate a glowing blue light.

There isn't much to see in terms of marine life. Bright green algae clings to the rocks, flowing out into the depths with hairlike strands (often referred to as "troll's hair"), and the occasional fish might flicker by, but Thingvellir's attraction is being the only place in the world you can dive between continental plates, which are moving apart at a rate of around two centimeters (0.8 inch) a year.

This national park and UNESCO World Heritage site is unique—culturally, historically, geologically, and diving-wise, a worthy addition to any logbook. ❍

"I watched my dive buddies ascend from the [Silfra] Cathedral, their bubbles forming sparkling white trails to the distant surface—a reminder that we are in fact underwater when the water is so clear it appears we can fly."

—JENNIFER ADLER OWEN, CONSERVATION PHOTOGRAPHER, CAVE DIVER, AND EDUCATOR

What You'll See: Two Continents • Clear Water • Algae • Trout (on occasion)

In crystal clear freshwater, a diver explores the fault lines of Silfra Canyon.

MEXICO

AFUERA

The largest aggregation of whale sharks in the world—and an underwater art gallery

AVERAGE WATER TEMP: 82°F (27.8°C) **AVERAGE VISIBILITY:** 20 to 100 feet (6 to 30 m)
AVERAGE DEPTH: 100 feet (30 m) **TYPE OF DIVE:** Open water, reef, and sculpture garden

Isla Mujeres, the Island of Women, is known for its spectacular sunsets, friendly locals, and white sand beaches. This five-mile-long (8 km) island is located off Mexico's Yucatán Peninsula, within spitting distance of Cancún. Once a former fishing village, it's now known for one thing: the largest aggregation of whale sharks in the world.

In Afuera, 25 miles (40 km) from Isla Mujeres, the world's biggest fish congregate to feed on fish eggs. Little tunny, a small tuna species that can produce up to 1.75 million eggs, spawn here, clouding the water with eggs and attracting gentle giants in staggering numbers. One aerial flight recorded 420 whale sharks in a single survey.

Although the experience is snorkel-only, it is unforgettable: Sharing the water with these distinctively patterned behemoths, 25 feet (7.6 m) in length with a gaping mouth five feet (1.5 m) across, gracefully weaving around swimmers as they go about their business of feeding, is both peaceful and exhilarating at the same time. Incredibly docile and agile, whale sharks are protected here by the establishment of the Whale Shark Biosphere Reserve in 2009, and whale shark tourism is highly regulated. (Do your part by following all guidelines, going out with licensed operators only, and doing your homework.) Unfortunately, Afuera is located just outside the protected area, but never underestimate the power of tourism: The more value placed on these magnificent creatures, the more likely

> *"We learn in school that whale sharks are* huge, *but that means nothing until you are . . . looking into the golf ball–size eyes of a whale shark the size of a school bus."*
>
> —JENNIFER ADLER OWEN, CONSERVATION PHOTOGRAPHER, CAVE DIVER, AND EDUCATOR

Life-size sculptures enthrall divers in the underwater museum at Manchones Reef.

Every summer, more than 300 whale sharks aggregate in the waters off Isla Mujeres.

What You'll See: Whale Sharks • Bottlenose Dolphins • Manta Rays • Parrotfish • Butterflyfish • Nassau Groupers • Angelfish • Moray Eels • Barracuda • Spotted Drum • Tawny Nurse Sharks

local and federal governments are to increase measures to protect them.

If by some chance you get bored of whale sharks, Isla Mujeres is a beginner diver's paradise, with plenty of shallow reefs to explore. Nearby is the Mesoamerican Barrier Reef, the largest barrier reef in the Western Hemisphere. Atlantis Reef, a 33-foot-deep (10 m) coral paradise, is home to a wide variety of colorful fish—angelfish, butterflyfish, and parrotfish, as well as larger species like Nassau groupers and tawny nurse sharks.

Manchones Reef is another popular spot in clear, calm waters. Its main point of interest is the 500 underwater sculptures created by environmentalist and underwater photographer Jason deCaires Taylor. Known collectively as MUSA (El Museo Subacuàtico de Arte), it gives divers the feeling of visiting a silent, underwater art gallery. The sculptures catalyze new coral growth and provide a secondary attraction for divers, to alleviate pressure on other reefs. ❍

Travel Tip:

June through September is the best time to see whale sharks, with peak times in July and August. Day trips are available from Cancún and Isla Mujeres. Whale shark tourism is highly regulated in Mexico, and following these strict rules could be the key to whale shark conservation.

ST. VINCENT AND THE GRENADINES

CRITTER CORNER

The macro capital of the Caribbean

AVERAGE WATER TEMP: 80°F (26.7°C) **AVERAGE VISIBILITY:** 25 to 150 feet (7.6 to 45.7 m) **AVERAGE DEPTH:** 32 feet (9.7 m) **TYPE OF DIVE:** Macro

Quality muck diving in the Caribbean—who knew?

Critter Corner is a shallow dive (just 32 feet/9.7 m) that stirs up some of the best macro life in the seas, a collection of exotic critters destined to drain the batteries of even the most jaded underwater photographers.

For the eagle-eyed who delight in detail, this sea grass bed near the southern tip of St. Vincent offers ample opportunity to spot longsnout seahorses, pipefish, yellowface pike blennies, and more.

Far from being a silty, low-vis experience, the volcanic black sand (courtesy of St. Vincent's still active volcano) falls heavily and quickly if it gets stirred up, providing a calm, clear viewing area for frogfish, jawfish, shrimp, and red-banded lobster.

Other nearby dive sites are worth a visit. Orca Point is a favorite for night divers seeking shrimp, scorpionfish, and decorator crabs. Anchor Reef offers regular sightings of moray eels and frogfish. New Guinea Reef is something a little different, with walls falling from 30 to 150 feet (9 to 45.7 m), complete with underhangs and gardens of black coral and sponges.

Most of St. Vincent's dive sites are located on the sheltered southwestern coast, blocking Atlantic swells and persistent easterly trade winds. The best time to visit is between May and November, especially if you're looking to avoid the crowds and peak rates from December through April. Keep in mind, hurricanes are possible between July and October. ○

What You'll See: Longsnout Seahorses • Pistol Shrimp • Yellowface Pike Blennies • Pipefish • Moray, Snake, Spoon-Nose, and Short-Tailed Eels • Frogfish • Jawfish • Shrimp • Octopuses • Butterflyfish • Flying Gurnards • Red-Banded Lobster

A secretary blenny curiously peers out from an encrusting sponge.

AUSTRALIA

COCOS (KEELING) ISLANDS

Paradise in the Indian Ocean

AVERAGE WATER TEMP: 82°F (27.8°C) **AVERAGE VISIBILITY:** 66 to 100 feet (20 to 30.5 m)
AVERAGE DEPTH: 30 to 130-plus feet (9 to 39.6+ m) **TYPE OF DIVE:** Shore, reef, drift, and wreck

The Cocos (Keeling) Islands (CKI) is the place where beginner divers can have the opportunity to be explorers.

This horseshoe coral atoll of 27 islands (only two of which are inhabited) is a remote speck in the vast Indian Ocean, closer to Jakarta, Indonesia (797 miles/1,282.6 km), than to Western Australia (1,736 miles/2,794 km). Its looks are the definition of paradise, with coconut palms, sugary white sand beaches, and pale turquoise lagoons, but its distance means it's not a well-known dive destination. Divers of any level can be among the first to explore this untouched and pristine area.

CKI has a wide variety of diving, including shore, reef, wreck, and drift dives. There are more than 30 popularly visited sites, with new areas regularly discovered. The islands are home to more than 500 species of fish, 600 species of mollusks, 200 species of crustaceans, and 100 species of hard corals. Sharks (whitetip reef, blacktip reef, and gray reef), turtles (hawksbill and green), dolphins (common and bottlenose), manta rays, and fish (triggerfish, butterflyfish, and angelfish) swarm the seas, and the outer lagoon is home to a beloved local dugong named Kat.

With wide-ranging visibility and water temperatures bordering on hot (77°F to 84°F/25° to 29°C), CKI is a prime paradise waiting to be explored and enjoyed. ❍

What You'll See: Hawksbill and Green Turtles • Bottlenose and Common Dolphins • Manta Rays • Cocos Pygmy Angelfish • Dugong • Gray Reef Sharks • Whitetip Reef Sharks • Blacktip Reef Sharks • Butterflyfish • Triggerfish • Surgeonfish

A shoal of yellowfin goatfish swim past a diver above a coral reef.

AUSTRALIA

PORT PHILLIP BAY

An underwater wilderness close to Australia's second largest city

AVERAGE WATER TEMP: 65°F (18°C) **AVERAGE VISIBILITY:** 50 feet (15 m)
AVERAGE DEPTH: 40 to 130 feet (12 to 39.6 m) **TYPE OF DIVE:** Open water and shore

Melbourne's Port Phillip Bay is a bustling port and recreational destination for Melbourne's 4.3 million residents (not to mention its annual influx of more than a million international visitors).

This bowl-shaped bay (actually made up of 16 bays) is cradled by the arm of the Mornington Peninsula reaching around until Portsea nearly touches Point Lonsdale. The depth fluctuates from 40 feet to 130 feet (12 to 39.6 m) here, holding a surprising variety of underwater wilderness areas.

Divers can access Port Phillip Bay by shore or boat, and underwater lies a wealth of places to explore—rock reefs, jetties, kelp forests, sponge gardens, sea grass beds, and wrecks. There is a spot to match every diver's level and interest.

Beginner divers enjoy the Octopus's Garden at Rye Pier, a 650-foot (198 m) swim, complete with informative underwater signage. Keep an eye out for pygmy seahorses, sponges, sea stars, and—of course—octopuses.

Flinders Pier is another easy shore dive, with a resident population of weedy seadragons, plus bull rays, nudibranchs, and the occasional visiting seal.

Looking for their next meal, Australian fur seals dive into the waters of Port Phillip Bay.

What You'll See: Cuttlefish • Giant Sea Stars • Giant Spider Crabs • Seahorses • Octopuses • Pygmy Seahorses • Pipefish • Weedy Seadragons • Bull Rays • Shipwrecks

Another favorite with novice divers: The Pope's Eye, an artificial reef built in the 1880s, rich with marine life. It was intended to become a fortress guarding the entrance to Port Phillip Bay, but was never completed, allowing critters and plants to make it their home, including kelp, nudibranchs, sea urchins, sea squirts, sea stars, and feather stars.

For divers wanting to explore a little farther and a little deeper, the vertical wall dives at Lonsdale Wall and Port Phillip Heads are popular, with more than 50 shipwrecks in the area to explore, including four submarines and the purposely sunk H.M.A.S. *Canberra,* a 445-foot (135.6 m) guided missile frigate lying in 98 feet (30 m) of water. You can even dive for your dinner (scallops, crayfish, and abalone) in Port Phillip Bay, provided you follow all rules and regulations.

But the area's real diving drawing card is the May through July congregation of giant spider crabs. These long-legged creatures aggregate in the shallow water around Port Phillip's beaches and piers, carpeting the sandy bottom—an insatiable, orange army scavenging whatever food it can find. This is one of the most unique spectacles in the marine world, a definite for the diving bucket list, and all within easy reach of one of the world's most vibrant and lively cities. ❍

"Melbourne is an easy-access place to discover the secret world of the blue-ringed octopus, striped cuttlefish, and weedy seadragon on patrol in the kelp beds. On the other side of the peninsula, at Phillip Island, little blue penguins called fairy penguins come ashore at nightfall."

—DAVID DOUBILET, UNDERWATER PHOTOGRAPHER

A venomous southern blue-ringed octopus emerges from the cracks of a sponge.

The southern blue devil is native to these Australian waters.

Knobby, or short-headed, seahorses hang on to sea grass.

SCUBAPRO

A diver breaks up an aggregation of spider crabs scavenging for food on the sandy bottom.

BONAIRE

SOMETHING SPECIAL

Night diving for beginners and a plethora of frogfish

AVERAGE WATER TEMP: 80°F (26.7°C) **AVERAGE VISIBILITY:** 75 feet (22.9 m)
AVERAGE DEPTH: 50 feet (15 m) **TYPE OF DIVE:** Open water, night, reef, wreck, and wall

Bonaire's license plates say it all: "Diver's paradise." This country established a marine park in 1979 that has allowed coral reefs and fish populations to flourish, and this offshore, natural aquarium, combined with a geography that's created unwinding miles of coastline sheltered from relentless trade winds, has made Bonaire a legendary diving destination.

Located north of Venezuela and southeast of Aruba, Bonaire features more than 100 dive sites accessible by both shore and boat: wrecks, reefs, walls, and drop-offs. Something Special is a site that lives up to its name, although the origin of the moniker remains somewhat dubious. (Stories tell of an infamous tryst or a cheap brand of rum; either way, the origin seems delightfully nefarious.)

Something Special is known for its multitude of frogfish and variety of macro life, not to mention angelfish, banded coral shrimp, squid, and eagle rays. It's easily reached by boat, but a shore entry over sand and debris is also possible.

This spot is really Something Special at night, when nocturnal critters like lobster, crab, and octopus make an appearance. It's an easy night dive for newbies, with little current and plenty to see.

To visit Something Special and the other 100 dive sites in the Bonaire National Marine Park, you will be required to pay a one-time entrance fee and attend an orientation with a dive operator. The orientation consists of dry and wet parts: a briefing on the marine park's rules and regulations, and a checkout dive (even for repeat divers). Trust us, it's worth the time and investment. ❍

What You'll See: Frogfish • Squid • Eagle Rays • Garden Eels • Angelfish • Banded Coral Shrimp • Green, Loggerhead, and Hawksbill Turtles • Seahorses • Octopuses

A goldentail moray eel comes out for its close-up.

MALDIVES

ARI ATOLL

Submerged pinnacles that are a pelagic heaven

AVERAGE WATER TEMP: 81°F (27°C) **AVERAGE VISIBILITY:** 75 feet (22.9 m)
AVERAGE DEPTH: 100 to 130 feet (30 to 39.6 m) **TYPE OF DIVE:** Open water

Oval-shaped Ari Atoll is a collection of 105 islands studding aquamarine seas. Its reputation as one of the best dive sites in the Maldives archipelago is due to its unique geography. Situated on top of an underwater mountain range, Ari Atoll is surrounded by submerged pinnacles, rather than a barrier reef. These pinnacles, and the channels running in between, are aggregation sites for large marine life, lending Ari Atoll its nickname, Pelagic Heaven. Divers come here for the big critters: whale sharks, manta rays, and schools of hammerheads.

Although the mantas and whale sharks can be found throughout the year at the Maldives, they move with the food, usually frequenting the plankton-rich waters of Ari Atoll February to May, although they sometimes arrive as early as December. (The South Ari Atoll Marine Protected Area, established in 2009 and the largest protected area in the Maldives at 10,378 acres/42 sq km, is one of the only places in the world where whale sharks consistently aggregate year-round.)

When these beautiful giants are absent, there is still a wealth of megafauna to marvel at, including hammerheads, gray reef sharks, shining trains of fusiliers, moray eels, blue line snapper, barracuda, guitarfish, batfish, and giant frogfish.

Because Ari Atoll isn't protected by walls and reef, dive sites are subjected to strong currents and varying conditions, which can push beginner divers to the edge of their comfort zones. Listen to the advice of dive guides and instructors, and dive within your limits and the conditions. With a large number of dive sites to choose from, it's likely you'll be able to find one that suits your skill level.

Maaya Thila is often called the most popular dive site in the Maldives. This pinnacle begins in the shallows (20 feet/6 m) and hosts frogfish, nudibranchs, and moray eels. As you descend to 40 feet (12 m), schools of fish flash past, along with soaring

Manta rays seem to fly through the waters of Ari Atoll.

Lucky snorkelers and swimmers look down on a curious whale shark.

What You'll See: Manta Rays • Whale Sharks • Gray Reef Sharks • Hammerheads • Triggerfish • Blueline Snapper • Batfish • Fusiliers • Napoleon Wrasses • Moray Eels • Barracuda

eagle rays, guitarfish, whitetip reef sharks, and gray reef sharks.

Hafsa Thila is another highlight in northern Ari Atoll, a pinnacle dive with a small, rounded, shallow top (40 feet/12 m) that spreads wider as you descend, like an underwater volcano. Nearby, Five Rocks is a circle of five jagged outcrops, with gorgonian- and anemone-filled channels, honeycombed rocks offering homes to mantis shrimp and moray eels, and swim-throughs, caverns, and overhangs.

Although Ari Atoll is diveable year-round, it is exposed and conditions vary with the season. December through March offers the best visibility, with top-of-the-line diving in February and March. April through June are the warmest months, and the wet season arrives May to August, bringing with it reduced visibility and increased wind and waves. ❍

Travel Tip:

Ari Atoll is 25 miles (40.2 km) long and the second largest atoll in the Maldives. It's accessible via seaplane, speedboat, or a local ferry. Many of the atoll's 105 islands have resorts, homestays, and B&Bs, although double-check if they offer a dive shop before booking. Liveaboards also abound and are an excellent way to experience a wide variety of the dive locations on offer.

GREECE

CHIOS ISLAND

A mind-boggling array of underwater caves, wrecks, and rock formations

AVERAGE WATER TEMP: 70°F (21°C) **AVERAGE VISIBILITY:** 20 to 165 feet (6 to 50 m)
AVERAGE DEPTH: Max 100 feet (30 m) **TYPE OF DIVE:** Open water, cave, and wreck

Chios is one of those places that's humming under the radar—so far. Located closer to the Turkish coast (just over four miles/6 km) than to Athens (130 miles/209 km), Greece's fifth largest island is keeping a remarkably low profile, maintaining its cultural color, known only to a few passing yachties who stay tight-lipped about what lies beneath.

Under the surface of the cool, electric blue Aegean Sea are shipwrecks straight out of storybooks. One wreck, lying below recreational dive limits, is from around 350 B.C. It was carrying amphorae, ceramic storage jugs containing wine and olive oil.

Rock formations, caves, and walls provide a dramatic background to any dive, especially the Great Wall, a 100-foot (30 m) vertical rock face studded with colorful coral. If you're lucky, tuna or swordfish might pass by, mixing with resident wrasses and perch.

Although Chios might sound otherworldly, it's a good spot for beginner divers, with moderate temperatures, a range of dive sites, and good visibility. (The best time of year to visit is April through October, when the water is warmest.) Here, you'll feel like you're diving through history, exploring paths winding through reefs and rocks that witnessed ancient voyages.

Tourism across the Greek Islands is holding steady, but be prepared for the unexpected on your travels here; there are often transportation and banking disruptions, among many other problems. Chios maintains an independent economy—thanks in part to its resident mastic trees that produce everything from chewing gum to toothpaste—so your visit to this island should be safe and simple. ❍

What You'll See: Dramatic Rock Formations (including walls and caves)
• Shipwrecks • Tuna • Swordfish • Wrasses • Perch

A church overlooks Chios's rocky shore.

GUAM

TOKAI MARU & S.M.S. CORMORAN

Dive wrecks from WWI and WWII share the same site.

AVERAGE WATER TEMP: 84°F (29°C) **AVERAGE VISIBILITY:** 50 feet (15 m)
AVERAGE DEPTH: 50 to 100 feet (15 to 30 m) **TYPE OF DIVE:** Wreck

It's not often you get to rub shoulders with history like this. Guam's wrecks have a spot on divers' bucket lists for two reasons: The first is diving two wars on one tank. Lying side by side in 120 feet (36.6 m) of water in Apra Harbor, the Japanese *Tokai Maru* from WWII and the German S.M.S. *Cormoran* from WWI rest in the same location, one of the only places in the world you can experience wrecks from two world wars.

The other reason is that Guam's wrecks are easily accessible to beginner divers. The *Tokai Maru* and S.M.S. *Cormoran* can be experienced by open water, advanced, and wreck divers alike. Another Apra Harbor favorite, a submerged American tanker—now a bustling reef resting in 50 feet (15 m) of water—is especially ideal for beginners.

Once outside the harbor, Guam's reefs and walls are popular spots to view the area's 1,000-plus species of fish, including reef sharks, tuna, giant trevallies, and lionfish, as well as spinner dolphins, pilot whales, green turtles, octopuses, and Christmas tree worms.

This American territory, the largest island in Micronesia, is an exciting destination for divers, one that is still mostly untouched, with a tropical and welcoming year-round climate. Located about 5,800 miles (9,334.2 km) west of San Francisco and 1,600 miles (2,575 km) east of the Philippines, Guam is reached by flights from Hawaii, Australia, and eastern Asia. U.S. citizens don't need a passport, but do need proof of citizenship (that is, a certified birth certificate). On land, don't miss the night markets, like Chamorro Village's Wednesday Night Market and the Mangilao Night Market, held most Thursday evenings. ❍

What You'll See: WWI and WWII Wrecks • Green Turtles • Pilot Whales • Spinner Dolphins • Lionfish • Christmas Tree Worms • Tuna • Octopuses • Giant Trevallies

A diver in Guam explores the remains of the *Tokai Maru*.

CALIFORNIA

SANTA CATALINA ISLAND

Explore otherworldly kelp forests teeming with life.

AVERAGE WATER TEMP: 62°F (16.7°C) **AVERAGE VISIBILITY:** 50 feet (15 m)
AVERAGE DEPTH: 20 to 100-plus feet (6 to 30+ m) **TYPE OF DIVE:** Shore

Twenty-two miles (35 km) from Los Angeles lies a world as vibrant, varied, and populous as the City of Angels. Easily accessible from the shores of Santa Catalina Island (better known as Catalina Island), underwater forests of kelp sway in filtered sunlight, looming large in the blue surge of the ocean.

Bright orange Garibaldi, a fearless damselfish and California's state fish, dart in and out of cover below sea otters resting on the surface. Huge black sea bass take their time weaving through thick fronds of kelp that can grow a staggering two feet (0.6 m) a day in the rich waters.

Everywhere you look, there is something to see. Everywhere you turn, there is life. Catalina Island is part of the Channel Islands, an eight-island archipelago off California's southern coast, and part of the 250,000-acre (1,012 sq km) Channel Islands National Park. Catalina receives nearly one million visitors a year (many via ferry), thanks to its proximity to Los Angeles, and many of those visitors are divers. That's because some of the best sights here lie under the water.

With more than 70 dive sites, Catalina has something to offer divers of every level and ability, and with easily accessible shore dives, it's a great spot for beginners to try their hands at cooler water and slightly surge-y diving.

The Casino Point Underwater Park is one of the most popular dive sites on the island, with steps leading directly into the ocean and depths ranging from 20 to 100 feet (6 to 30 m). Established in 1962, this no-take (no fishing, shellfish gathering, or salvaging wrecks allowed) underwater park is 2.5 acres (.01 sq km) in size, with plenty to explore. Along with kelp forests, there are also a few small wrecks to look at (*Kismet* and *Sue-Jac* are just a short swim from shore), plus plenty of marine life, like octopuses, Garibaldi, abalone, and nudibranchs.

Kelp forests shelter the colorful coral
beneath the waters of Catalina.

A Garibaldi fish and giant spined sea star find homes in the kelp forests.

What You'll See: Kelp Forests • Giant Black Sea Bass • Sea Otters • Sea Lions • Garibaldi Damselfish • Nudibranchs • Zebra Sharks • Halibut • Guitarfish • Lobsters • Abalone • Mantis Shrimp • Moray Eels

The Blue Cavern is another pristine area, a strictly monitored preserve with a wall drop-off from 30 feet to 75 feet (9 to 22.9 m), hanging sea fans, and caves. A plethora of creatures, including kelp, nudibranchs, moray eels, gobies, and butterflyfish gather at the Sea Fan Grotto, thanks to a consistent current carrying nonstop fast food for fish.

Ship Rock is another diver favorite, rising 120 to 200 feet (40 to 67 m) from the sand to above the surface. Divers who have been report it feels like a new dive every outing. Because it sits offshore, Ship Rock attracts everything from seals to nudibranchs, and is always full of surprises.

Time between dives? Catalina has plenty to offer. This compact island (22 miles/35 km long and eight miles/13 km across) has been inhabited for 7,000 years by smugglers, missionaries, pirates, gold diggers, film stars, and even the Union Army. Today it is home to 4,000 permanent residents above water, and a rich underwater world teeming with life just offshore. ❍

Travel Tip:

Catalina is a year-round dive destination, but visibility can vary with the season. It's best from September to November. April and June can have lowered visibility due to plankton blooms. Thermoclines are consistently present around 50 feet (15 m).

BRITISH VIRGIN ISLANDS

R.M.S. *RHONE*

Dive the haunted wreck of a ship bested by a hurricane.

AVERAGE WATER TEMP: 80°F (26.7°C) **AVERAGE VISIBILITY:** 40 to 100 feet (12 to 30 m)
AVERAGE DEPTH: 35 to 80 feet (10.7 to 24 m) **TYPE OF DIVE:** Wreck

In 1867, Captain Woolley steered the 310-foot (94.5 m), iron-hulled *Rhone* into the San Narciso hurricane. Legend has it that the captain, believing the eye of the storm was its end, celebrated prematurely with a cup of hot tea on the deck. The hurricane resumed and tossed him overboard. Another legend holds the teaspoon embedded in the coral was his.

When the storm threw the *Rhone* onto Black Rock Point, the hull breached, and the cold seawater collided with the overheated boilers, resulting in a resounding explosion that tore the ship in half. Because most of the passengers had been tied to their bunks to keep them secure (preventing injuries and panic), the survivor count was low.

The *Rhone* lies in its two parts on a site that became the British Virgin Islands' first national marine park (roughly 800 acres/3.2 sq km) in 1980. It's recommended to do two dives: one for the shallow stern, complete with an original porthole and one of the oldest brass propellers in the world, and one for the bow, lying 100 feet (30 m) from the stern in deeper water, complete with a foremast that still has the crow's nest attached.

On the bow side, the ship's interior is easily accessible, encrusted with coral and sponges, and home to yellowtail snapper and barracuda. The stern isn't in as good condition, but there's a short swim-through near the propeller and plenty of fish to spot, including pufferfish, moray eels, and sergeant majors.

The ship's tragic past has led to more than one haunted tale—divers claim to see swimmers (with no gear) frantically clawing their way to the surface; feel invisible hands on their shoulders; or hear strange calls and underwater screams. ❍

What You'll See: Barracuda • Stingrays • Turtles • Green Morays • Yellowtail Snappers • Jacks • Orange Cup Coral • Sponges • Tarpon • Arrow Crabs • Lobster • Sergeant Majors • Pufferfish • Damselfish

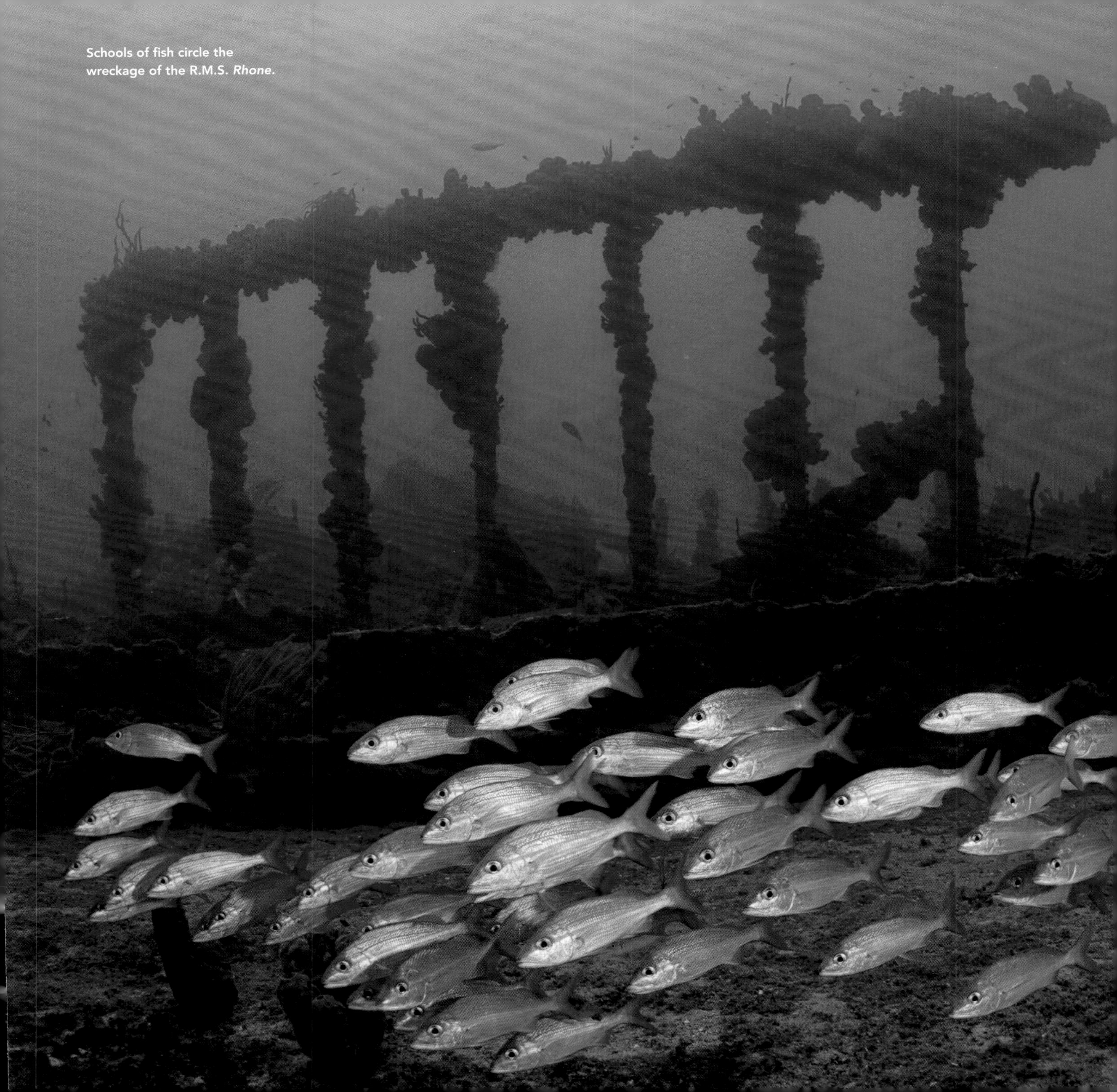

Schools of fish circle the wreckage of the R.M.S. *Rhone*.

THAILAND

RICHELIEU ROCK

Thailand's most famous dive site has everything from harlequin shrimp to whale sharks.

AVERAGE WATER TEMP: 81°F (27°C) **AVERAGE VISIBILITY:** 49 to 115 feet (15 to 35 m)
AVERAGE DEPTH: 16 to 115 feet (4.9 to 35 m) **TYPE OF DIVE:** Open water

Richelieu Rock is a tiny hunk of rock in the open ocean, named for a U.S. Navy captain who had the misfortune of running his destroyer into it during WWII. Located northeast of the Similan Islands (page 336) in the emerald Andaman Sea, this is Thailand's most famous dive site, and with good reason.

Richelieu Rock teems with life, from the small (harlequin shrimp, seahorses, ghost pipefish, and nudibranchs), to the large (moray eels, cuttlefish, and barracuda), to the huge (manta rays and whale sharks), to the rare (a species of tomato clownfish native to this site). Sea fans sway in nutrient-rich waters, and it's well worth carrying a small torch to peer under overhangs—given the small amount of real estate at Richelieu, creatures occupy every nook and cranny.

This horseshoe-shaped site changes with the seasons. It's accessible from October until early May, due to its open ocean location. February to April is the best time for manta rays and whale sharks, with March and April having lower visibility due to plankton-rich waters. Lunar cycles and thermoclines can also wreak havoc on visibility and water temperature.

Richelieu Rock is best dived via liveaboard. Day trips are available, but they are long days and uncomfortable rides. Do your homework and nab an experienced guide—there are a lot of hidden treasures at Richelieu, and having a guide who knows where to look is golden. This spot can be crowded, so be prepared, dive with patience, and always use a surface marker buoy (SMB) while surfacing due to boat traffic. ❍

What You'll See: White-Mouth Moray Eels • Pipefish • Sea Fans • Whale Sharks • Manta Rays • Anemones • Barrel Sponges • Barracuda • Cuttlefish • Seahorses • Frogfish • Octopuses • Scorpionfish • Trevallies • Harlequin Shrimp • Groupers

A day octopus is almost lost in the brilliant coral colors.

AUSTRALIA

PORT LINCOLN

The only place in the world to ocean-floor cage dive with great whites

AVERAGE WATER TEMP: 63°F (17°C) **AVERAGE VISIBILITY:** 33 feet (10 m)
AVERAGE DEPTH: 82 feet (25 m) **TYPE OF DIVE:** Cage

You strain your eyes, looking for any flash of white. Even with good visibility, great white sharks (*Carcharodon carcharias,* "the ragged-tooth one") are remarkably adept at blending in, with specific coloring that helps them vanish in a few feet of water. Perhaps that's why the world's largest predatory fish has survived for millions of years. Perhaps that's also why, after coexisting with these charismatic creatures for hundreds of thousands of years, we still know so little about them.

If you want to learn more, Port Lincoln is the place to get educated. This sharky town has three shark-cage diving operators: Two offer surface-cage diving, while the third is the only operator in the world to offer ocean-floor cage diving with great white sharks for scuba-certified divers.

Three divers and one crew member with a dive master qualification—one for each corner of the cage—are lowered 60 to 80 feet (18 to 24 m) into an underwater world that is the sharks' domain. Down here, great white sharks behave differently than they do on the surface. Closer to the surface, the sharks are more alert. In this bottom world, with

"Great white shark diving started in Port Lincoln with Rodney Fox, a shark attack survivor who wanted to safely introduce others to this majestic creature. [Port Lincoln] is now a center of gravity for great white enthusiasts and a special place to encounter the curious nature of the endangered Australian sea lions at Hopkins Island."

—DAVID DOUBILET, UNDERWATER PHOTOGRAPHER

Closer to the ocean floor, a great white calmly cruises the waters.

The rocky coastline of the North Neptune Islands

What You'll See: Great White Sharks • Bronze Whaler Sharks • Mako Sharks • Blue Sharks • Thresher Sharks • Epaulette Sharks • Orcas • New Zealand Fur Seals • Australian Sea Lions • Kingfish • Smooth Stingrays • Southern Eagle Rays • Horseshoe Leatherjackets • Trevallies

swaying sea grasses, and smooth stingrays and southern eagle rays flying over sandy patches on the ocean floor, the great whites are more relaxed, more curious. It's not unusual for divers to see three or four different great whites from the ocean-floor cage on a single dive, while surface cage divers see one or two. (The record for the ocean-floor cage is 19 individual great whites in a 15-minute period—ironically while a new shark repellent device was being tested.)

With more than 1,000 individual great white sharks having been identified in the Neptune Islands in the past 18 years, shark research is a significant focus for the operator offering expedition-style, multiday liveaboard trips, which feature a couple of daily dives (either ocean-floor or surface cage, sometimes both), along with informative shark talks in the evening.

Travel Tip:

Expeditions take place year-round. Shark numbers increase in summer, but mainly with smaller males. For the big super-females (reaching 19 feet/5.8 m), brave the colder and rougher weather in June and July. The Neptune Islands are a three- to five-hour boat ride from Port Lincoln. The seas can roll at any time of year, so seasickness tablets are recommended for the journey. Don't miss the opportunity to snorkel with the sea lions off Hopkins Island.

A pair of sea lions play close to the water's surface.

A male great white cruises by a school of silver trevallies.

The sharks have been sighted at the Neptune Islands since explorer Matthew Flinders first named the islands in 1802. His impression of these two low hills of rock and scrub—North and South Neptune—was one of a remote and inaccessible place. More than 200 years later, these granite islands are much the same, protected in a marine park, home to Australia's largest colony of New Zealand fur seals.

Although two of the three cage diving operators have been granted licenses to use berley (chum) to attract sharks to the boats, the licenses come with tight rules to protect the sharks and regulate the industry, giving divers peace of mind that good practice is being followed.

Ocean-floor cage diving with great white sharks is a mesmerizing and wonder-filled experience that is not to be missed. Enclosed, with no worries about buoyancy, divers have nothing to do but marvel at these extraordinary creatures. ❍

SOUTH AUSTRALIA

RAPID BAY JETTY

The best place in the world to spot leafy seadragons

AVERAGE WATER TEMP: 62°F (16.7°C) **AVERAGE VISIBILITY:** 15 to 50 feet (4.6 to 15 m)
AVERAGE DEPTH: 22 to 40 feet (6.7 to 12 m) **TYPE OF DIVE:** Jetty

Leafy seadragons have captured the heart of many a diver. These ornate and outlandish creatures are the stuff of dreams, with fluttering, leaflike appendages; a long, undulating body; and a wise expression in the eye. They are perfectly camouflaged in yellows, olives, greens, and browns to blend in with kelp and seaweed. They are one of the ocean's craziest, phantasmagoric creations, a rare and protected species found only in South Australia, Western Australia, and Victoria.

South Australia's Rapid Bay Jetty, 65 miles (105 km) south of Adelaide, is the most reliable place in the world to spy a leafy. They make their home under the old jetty, a T-shaped structure built in 1940 that is now closed to the public. (A newer, shorter jetty was built next to it. This jetty is open to the public and is the access point for divers.)

Navigation can be tricky on this dive—although it's a jetty dive, the two structures cast weird, disorienting shapes, and compasses are foiled by the metal construction. However, it's open year-round, and a variety of dive shops provide guided expeditions.

The old jetty is a haven for nudibranchs, leatherjackets, and pufferfish, with larger schools hanging out by the T. Even a few local bull rays surprise divers with their size and inquisitiveness. The show stealers, though, are the leafies.

Leafy seadragons are large (up to 14 inches/35.6 cm) and easy to spot, but they are easily stressed out. These protected creatures have a limited home range of about 30 feet (9 m)—pursuing them out of this region can distress them. Feel free to take photographs, but then back off for 15 minutes. Don't worry—leafies travel at a speed of an eighth of a mile an hour (0.2 km/hr), so you're not going to lose them. ❍

What You'll See: Leafy Seadragons • Weedy and Common Seadragons • Bull Rays • Feather Duster Worms • Colorful Sponges • Leatherjackets

The gem of Rapid Bay Jetty:
a leafy seadragon

MAURITIUS

TROU AUX BICHES

A diverse range of dive sites close to shore

AVERAGE WATER TEMP: 78°F (25.6°C) **AVERAGE VISIBILITY:** 50 feet (15 m)
AVERAGE DEPTH: 23 to 130 feet (7 to 39.6 m) **TYPE OF DIVE:** Drift, cavern, open water, and reef

Mark Twain wrote that Mauritius was made first, and then heaven, and heaven was copied after Mauritius. Twain was quoting a local from his visit to Mauritius in 1896, and the postcard-perfect looks of this Indian Ocean island nation certainly back up their opinions.

Located 1,242 miles (1,999 km) off the East African coast, this small island (40 miles/64 km long and 28 miles/45 km wide) has an enviable 200 miles (322 km) of coastline surrounded by an even more enviable shallow lagoon reef, ideal for beginner divers.

The surprising thing about Mauritius is the variety of diving available, both inside and outside of the lagoon. On offer: reef dives, coral arches, wreck dives (more than 20 from the 18th and 19th centuries, plus a few recently sunk ships), drift dives, and caverns. Schools of fish—barracuda, sweetlips, parrotfish—flash past in bright walls of moving color, while moray eels and octopuses keep a watchful eye on any diver finning past their hiding places.

Trou aux Biches, tucked away on Mauritius's northern coast, is home to a large number of the most easily accessible dive sites, providing new divers the opportunity to add several notches to their weight belts. These include the Aquarium (a sheltered dive with rich light and plenty of marine life, from hard and soft coral and nudibranchs to clownfish and parrotfish), Mimi's Arch (a small drop-off with a vast arch and swarming tropical fish like pufferfish and butterflyfish), and the wreck of *Stella Maru* (a Japanese freighter in 80 feet/24 m of water, home to a variety of life like stingrays and moray eels). ❍

What You'll See: Octopuses • Moray Eels • Clownfish • Parrotfish • Leaf Fish • Sea Grass Gardens • Bull Sharks • Whitetip Reef Sharks • Barracuda • Green and Hawksbill Turtles • Sweetlips • Pufferfish • Trumpetfish • Anemones

A bright orange school of crescent-tail bigeye lead the way for a diver.

JAPAN

YONAGUNI JIMA

Explore the mystery of the Underwater Monument.

AVERAGE WATER TEMP: 79°F (26°C) **AVERAGE VISIBILITY:** 180 feet (55 m)
AVERAGE DEPTH: 18 feet (5.5 m) **TYPE OF DIVE:** Open water

Mysteries still exist. Under the fabled clear water of Yonaguni Jima (reportedly one of the world's top five for visibility), a large monolith lurks. It covers an area of more than 11 acres (.04 sq km), a rectangular formation stretching 90 feet (27 m) toward the surface of the Pacific Ocean, lying only 16 feet (4.9 m) deep.

As divers fin through the ever present surge, they spy shapes resembling steps, pillars, columns, walls, arches, ledges, and pyramids. It is a journey that is both mesmerizing and mind-bending. Is this structure a natural geological occurrence? Or a Japanese Atlantis, an ancient man-made structure sunk by time? Therein lies the mystery. No one knows.

Discovered by local diver Kihachiro Aratake in 1986, the site has been combed over by thousands of underwater Indiana Joneses, looking for the code to crack the cipher. The decision is still split.

Some experts believe the formation is consistent with its sandstone and mudstone geology, which produces ruler-sharp edges. Any holes or bowls in the rock could be formed by underwater eddies working away at the stone, or marine creatures burrowing into seams. Naturalists claim that the symmetry doesn't add up as perfectly as it looks to the masked eye.

Other experts strongly disagree. They believe the stone structure might be the ruins of a 5,000-year-old city, including a castle, arch, temples, and stadium, connected by roads and framed by walls.

Japan has a history of seismic activity, including the unfortunate world record of the largest tsunami, which struck Yonaguni Jima in 1771, raging more than 130 feet (39.6 m) high. The "cityists" believe it wouldn't be a stretch to imagine that a similar event might have sent the city to its watery grave.

No direct human evidence—pottery, for example—has been found at the site, but there

The view from the cliffs of
Okinawa's Yonaguni Island

Divers explore the mysterious levels of the Yonaguni Monument.

What You'll See: Hammerheads • Rock Formations • Blue-Banded Snapper • Cardinalfish • Barracuda • Bigeye Trevallies • Dogtooth Tuna • Cuttlefish • Bannerfish • Sea Goldies • Boxfish • Pufferfish • Butterflyfish • Clownfish • Moray Eels • Octopuses

is a painted relief resembling a cow. (The naturalists believe these are just coincidental scratches.)

Either way, this unverified site is beautiful to behold and a modern-day mystery waiting to be solved—or perhaps marveled at simply the way it is. And even if no human relics are to be found, you can still spot plenty of marine life exploring the structure, including snapper, barracuda, trevally, and octopus.

Located at the southern tip of Japan's Ryukyu archipelago, about 75 miles (121 km) off Taiwan's eastern coast, the ruins (also known as the Underwater Monument) are a 15-minute boat ride from Yonaguni Harbor. Although the visibility is world-class, the currents can be a bear. It's an easy dive if the conditions are right (although surge is to be expected), so beginner or anxious divers should plan their dives for optimal days.

With nearly 70 documented dive sites in the area, and visiting schools of hammerheads, Yonaguni is a true wonder of the dive world. ❍

Travel Tip:

Yonaguni Jima is located in the Okinawa Prefecture of Japan, in the Pacific between Taiwan and mainland Japan. More Kauai than Kyoto, Okinawa is at around the same latitude as Hawaii and Florida, serviced by regular flights from Japan and China. The hammerheads visit between December to May, and typhoons can occur June through November.

COMOROS

MOHÉLI MARINE PARK

Beautiful diving paired with adventure travel—be one of the first to dive the "pearl of the Comoros."

AVERAGE WATER TEMP: 77°F (25°C) **AVERAGE VISIBILITY:** 75 feet (22.9 m)
AVERAGE DEPTH: 15 to 75 feet (4.6 to 22.9 m) **TYPE OF DIVE:** Open water

The dives are the easiest thing about the Comoros. Getting there is an adventure.

There aren't many places that are still "off the map," but this island nation in the Indian Ocean, located in the Mozambique Channel between Madagascar and Mozambique, is a pristine spot with large, healthy coral structures; rays; and reef fish.

Made up of four major islands, the Comoros have seen more than their fair share of political upheaval, and the country is a poor one. Despite this, it established the Mohéli Marine Park off the south coast of Mohéli, the smallest of the four islands, in 1998. Known as the "pearl of the Comoros," Mohéli is untouched, wild, and sparsely populated.

The reserve spans nearly 155 square miles (401 sq km) of coral reef and sandbanks—and is still relatively unexplored, a treat for divers looking for real adventure. Here you'll find pristine coral teeming with colorful fish (rays, reef sharks, clownfish, and barracuda, to name a few), turtles nesting on the beach, and hatchlings heading into the waves at dawn.

The Comoros Islands are some of the only places in the world where it is possible to see coelacanth, a strange, spotted fish with a seemingly excessive number of fins, once believed to have been extinct for millions of years. In the second half of last century, a scientist discovered Comorian fishermen regularly caught this elusive fish in deep waters.

Although populated by friendly people, the islands are still a politically unstable area—and access to Internet and electricity can be limited, so know the lay of the land before you go. Travel disruptions are the only constant. Bring as much of your own gear as you can, as lodge selections are limited. The nearest hyperbaric chamber isn't close, and a laundry list of vaccinations are advised, including hepatitis, rabies, tetanus, and typhoid. ❍

What You'll See: Coelacanth • Humpback Whales • Spotted Eagle Rays • Manta Rays • Barracuda • Gray Reef Sharks • Barrel Sponges • Electric Rays • Anemones • Clownfish

Barely visible, a shrimp hides inside a blue sponge.

MEXICO

DISCOVERY BAY

Warm(er) water cage diving with majestic great whites

AVERAGE WATER TEMP: 68°F (20°C) **AVERAGE VISIBILITY:** 125 feet (38 m)
AVERAGE DEPTH: 33 feet (10 m) **TYPE OF DIVE:** Cage

Isla Guadalupe is a lonely hunk of rock in the Pacific Ocean. Only 90 square miles (233 sq km) in size, it's home to approximately 200 fishermen, farmers, scientists, and military personnel—and roughly the same number of uniquely identified visiting great white sharks that travel to the area every August through November. Scientists believe the sharks make regular pilgrimages to feed on tuna and the Guadalupe fur seals that attracted Russian and U.S. hunters in the late 18th century.

One of the five countries in the world to cage dive with great whites, Guadalupe is unique in that its (relatively) warm water and consistently clear visibility means that divers get an eyeful of these 10- to 20-foot (3 to 6 m) charismatic giants.

A suspended, submersible cage holds four wide-eyed divers breathing from a hookah system (surface-supplied air), soaking in every moment with the ocean's undisputed apex predator gliding through a brilliant blue background. Seeing a great white shark in its territory is spine-tingling and unforgettable.

Guadalupe is only accessible via a multiday liveaboard, and there is little else to do (and little else you'll want to do) besides watch sharks and cycle in and out of the cage. Liveaboards run August through November, and the trip is a long one—18 to 22 hours departing from either San Diego, California, or Ensenada, Baja California, Mexico. ○

"To see great whites for the first time or the five hundredth time, the charm never wears off."

—BRIAN SKERRY

What You'll See: Great White Sharks • Guadalupe Fur Seals • Elephant Seals • Tuna (on occasion)

Cage divers experience a close encounter with a great white.

AUSTRALIA

JULIAN ROCKS

One of Australia's best and most underrated dive spots is located in a cool little town.

AVERAGE WATER TEMP: 71°F (21.7°C) **AVERAGE VISIBILITY:** 16 to 100 feet (4.9 to 30 m)
AVERAGE DEPTH: 78 feet (23.8 m) **TYPE OF DIVE:** Open water

Byron Bay has it going on. This laid-back town (population around 9,000) is located on Australia's most easterly point, approximately 110 miles (177 km) south of Brisbane and 500 miles (805 km) north of Sydney. It's known for its golden surf beaches, outdoor markets and music festivals, lush hinterland, yoga retreats, and a relaxed Aussie lifestyle Lord Byron would approve of. (Byron Bay was named after his grandfather, Admiral Byron, by Captain James Cook in 1770.)

Byron Bay is well known to travelers. What's less well known, however, is the pelagic party taking place a mile (1.6 km) offshore at a cluster of granite known as Julian Rocks. Twenty million years ago, a volcano blew its top, and Julian Rocks was born.

The rocky underwater terrain, easily accessed by a 10-minute boat ride, is filled with trenches, caves, pinnacles, and walls. A protected marine reserve since 1982, Julian Rocks is stationed in the middle of the East Australian Current, and its easterly location means that any creature migrating north or south along Australia's east coast (both tropical and cold-water species) stops by "the rock" for a visit.

Adding to the color of Julian Rocks, orange cup coral polyps extend to feed.

What You'll See: Zebra, Sand Tiger, and Wobbegong Sharks • Humpback Whales • Loggerhead, Hawksbill, and Green Turtles • Yellowtail Kingfish • Barracuda • Mulloway • Bull Rays • Nudibranchs • Bullseyes • Sweetlips • Cuttlefish

In the summer, that means manta rays and a beautiful collection of zebra sharks. In winter, sand tiger sharks and humpback whales are regular visitors.

Outside of seasonal visitors are plenty of year-round residents: Cuttlefish, sweetlips, Queensland groupers, bullseyes, nudibranchs, and turtles (hawksbill, green, and loggerhead) are frequently spotted, and wobbegong sharks gather by the hundreds, carpeting the bottom, the coral, and the rocks. In total, this hot spot supports more than 1,000 marine species, including 500 species of fish, as well as a variety of coral (even black coral).

Ranked one of the world's 10 best places to learn to dive, the variety of dive sites and depths, and the plethora of marine life paired with quality operators, makes Julian Rocks a great spot for beginners. The ever present current works to divers' advantage, carrying you around the shallow side of the rock, or the deeper side, depending on your choice. (And the entry and exit points are usually current free.) This is one of the only places in the world new divers can get up close and personal with sharks, whales, and rays—there's always something to see, and every dive is different. And if the conditions kick up, which they can, Byron is an easy place to wait out the weather, with its lively culture, fantastic food, and limitless hiking. ❍

"My first open water dive was at Julian Rocks. The first animal I saw was a sand tiger shark. My very first thought was: I'm going to spend my life diving. Years later, Julian Rocks is still my favorite place to dive because there is a huge amount of varied marine life—including humpback whales singing so loudly they're deafening."

—CHRIS TAYLOR, AUTHOR AND DIVE MASTER

An aerial view of Byron Bay

Above the water, visitors can whale watch from boats.

A porcupinefish explores the waters of Byron Bay.

A pair of bottom-dwelling wobbegong sharks rest below a school of fish and divers.

NORWAY

LOFOTEN ISLANDS

Swim in chilling seas with wild orcas.

AVERAGE WATER TEMP: 36°F (2.2°C) **AVERAGE VISIBILITY:** 50 feet (15 m)
AVERAGE DEPTH: 65 to 130 feet (19.8 to 39.6 m) **TYPE OF DIVE:** Open water snorkel

North of the Arctic Circle, there be giants. Orcas, with their distinctive black-and-white coloring, hold as much fascination for ocean lovers as great white sharks. Weighing up to six tons (5.4 metric tons) and growing up to 32 feet (9.8 m) in length, the largest of the dolphin family, and one of the world's most powerful predators, killer whales reign supreme in the cold waters of our planet.

And in the cold, clear waters of the Norwegian Sea, near the picturesque and colorful Lofoten Islands, divers can swim with these alluring and intelligent animals—on their terms.

A multiday liveaboard takes divers into the Norwegian Sea to look for the resident pods that make their home near the Lofoten Islands. The orcas themselves determine whether or not they want to interact with swimmers, but even if they have other business to attend to, watching the orcas from the boat, and seeing the triangular, six-foot-tall (1.8 m) dorsal fin of a large adult male slice the sea like a scythe, is worth the trip alone. In the wild, these creatures are extraordinary—graceful, powerful, and everything we want our natural and untamed oceans to be.

You aren't allowed to tank-dive with the orcas (snorkel or free-dive only—beginner divers should be comfortable in an open ocean with these giants), but as regulations in North America prohibit any swimming with orcas, and resident pods are difficult to come by, this unique experience is worth a mention. Norway also has plenty of diving to stave off cravings, including drift dives past kelp forests, strange-looking rock walls, and diving WWII German shipwrecks, well preserved thanks to the cold water. ❍

What You'll See: Orcas • Brown Algae Forests • Kelp Forests • Sea Stars • Anemones • Wolffish

The frigid waters are worth braving for a once-in-a-lifetime chance to swim with orcas.

SEYCHELLES

ALPHONSE ISLAND

The Eden of the Indian Ocean: verdant, pristine, and isolated

AVERAGE WATER TEMP: 84°F (28.9°C) **AVERAGE VISIBILITY:** 50 to 100 feet (15 to 30 m)
AVERAGE DEPTH: Up to 130 feet (39.6 m) **TYPE OF DIVE:** Open water and reef

The Seychelles, an archipelago off the coast of Africa in the Indian Ocean, are often referred to as the garden of the ocean, and in that garden, Alphonse Island is the Eden: pure, untouched, a raw and beautiful natural sanctuary.

This verdant, arrowhead-shaped island, carved in half by a landing strip, is scalloped by white beaches and populated with tortoises, crabs, and birds. Its isolation and lack of fishing make for uncrowded and interesting diving, with more than half of its 30-plus dive sites suitable for beginners. The aptly named Wonderland is a shallow (15- to 50-foot/4.6 to 15 m), low-current site with coral heads and overhanging reefs protecting lobsters, tasseled scorpionfish, and moray eels. Titan triggerfish, turtles (hawksbill and green), and snapper also like this sheltered spot, as do sharks. Tawny nurse, silvertip, gray reef, and whitetip reef sharks can all be seen in the turquoise waters, along with the occasional hammerhead.

Turtle Parade is another favorite—this plateau dropping down to a 100-foot (30 m) wall is home to its namesake turtles, as well as sweetlips, groupers, purple-tinged sea fans, unicorn fish, and surgeonfish.

Even House Reef, the shallow (20-foot/6 m) lagoon in front of Alphonse's dive center, is a unique place to explore, with rich corals (branching, plate, boulder, and mushroom), bright reef fish (butterflyfish, damselfish), and even a few bigger species (groupers, bluefin trevallies, and surgeonfish) cruising the outer wall. ❍

What You'll See: Green and Hawksbill Turtles • Moray Eels • Barracuda • Bigeye Trevallies • Batfish • Fusiliers • Snapper • Potato Cod • Groupers • Hammerheads • Bull Sharks • Silvertip Sharks • Tawny Nurse Sharks • Gray Reef Sharks • Whitetip Reef Sharks • Nudibranchs • Whip Coral Gobies • Peacock-Tail Anemone Shrimp • Tiger Mantis Shrimp • Pipefish

One of the three Alphonse atolls, St. Francois Island is an untouched Eden.

FIJI

BEQA LAGOON

Dive with more than eight species of sharks.

AVERAGE WATER TEMP: 78°F (25.6°C) **AVERAGE VISIBILITY:** 60 feet (18 m)
AVERAGE DEPTH: 80 feet (24 m) **TYPE OF DIVE:** Shark

In Beqa Lagoon, there's a clearly defined pecking order. Usually the tawny nurse sharks approach first, half a dozen deep, prompt and purposeful. Next, a lemon shark or two passes by. When the big bull sharks arrive, the silvertips get twitchy and zip in and out of view. Occasionally a tiger shark, beautifully patterned, 14 feet (4.3 m) in length, will make an appearance, and the other sharks give way, letting the tiger cruise its blue domain, mildly curious, eyes roving over the divers that are kneeling below.

Situated off Pacific Harbour on Viti Levu, Fiji's main island, Beqa (pronounced Benga) isn't the typical Fijian dive: Known as the Soft Coral Capital of the World, this Pacific Ocean nation, a collection of 332 islands (only 110 of which are inhabited) 1,100 miles (1,770 km) north of New Zealand, is legendary for its colorful coral gardens (390 species) and variety of fish (1,200 species). There are better places in Fiji than Beqa to see the coral wonder world, but Beqa holds the marine ace: sharks. Beqa Lagoon is one of the best shark dives in the world, a unique place where divers can see more than eight shark species in a single dive.

The enormous Beqa Lagoon covers more than 150 square miles (388.5 sq km). It's believed that more than 400 bull sharks—a territorial species, unlike their transient great white brethren—are local residents, which is part of what makes the shark diving here so successful: The sharks, shark handlers, and divers all know the drill, and everyone follows the rules.

Divers descend 80 feet (24 m) to a landing area known as the Arena. This isn't a cage, but the perimeter of a dive site, with a clear view of a metal box filled with fish scraps from a local seafood factory. The handlers are there to remind sharks of the routine, using staffs to (respectfully) push them out of the way if they get too close to divers.

Beqa Lagoon teems with life, including blacktip reef sharks and scissor-tail sergeant majors.

Encounter more than eight species of sharks on your dive here.

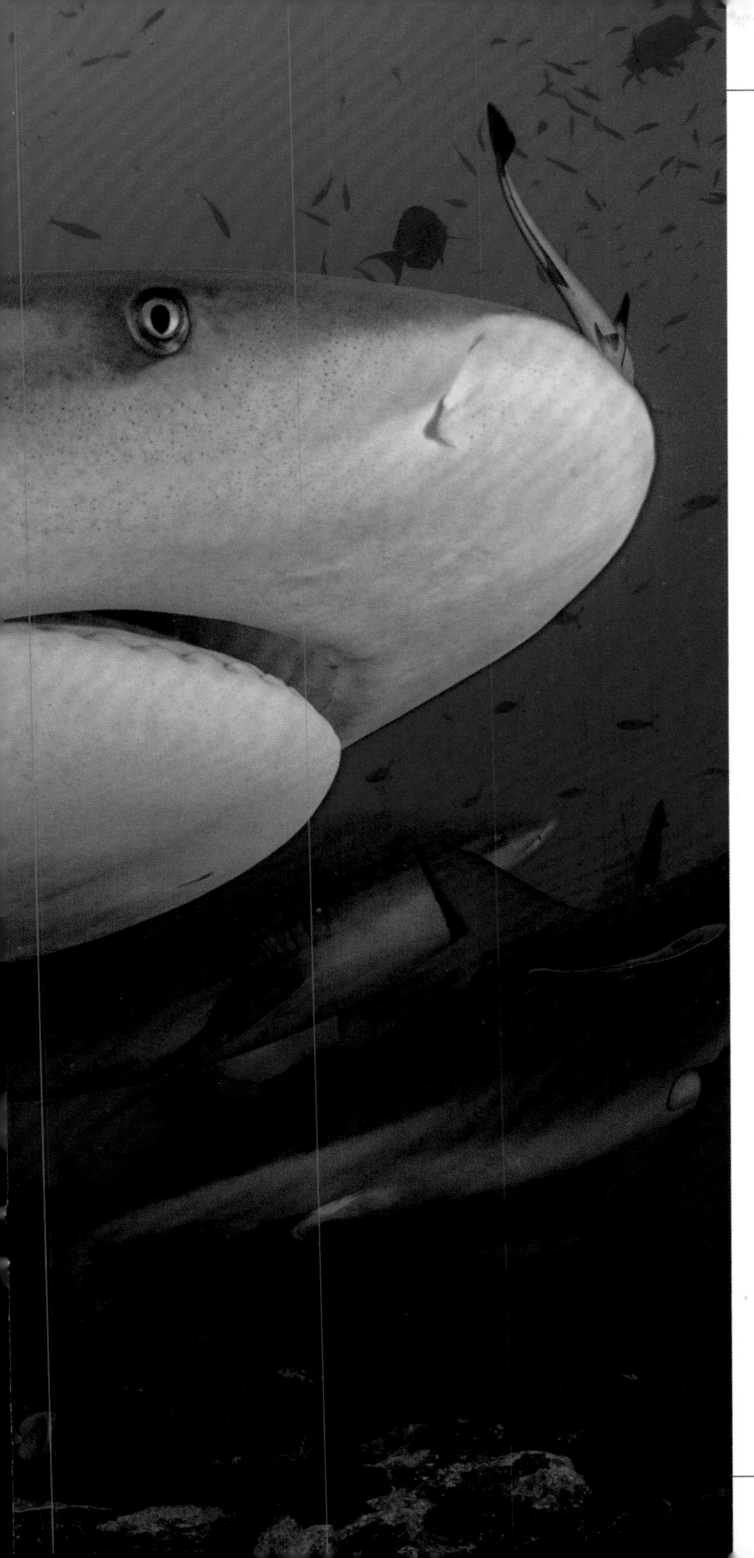

What You'll See: Bull Sharks • Tiger Sharks • Whitetip Reef Sharks • Blacktip Reef Sharks • Tawny Nurse Sharks • Lemon Sharks • Gray Reef Sharks • Silvertip Sharks • Humphead Wrasses • Queensland Groupers • Barracuda • Butterflyfish • Triggerfish

Sharks weave in and out of the Arena, queuing for their turn near the metal box as a cloud of fish block out the sun. The result is a finny maelstrom of flashing silver, white, and black, with the sharks' distinctive outlines reigning supreme, carving through the blue water. Divers travel to Fiji for the specific purpose of seeing this breathtaking spectacle.

The dives were started to educate divers about sharks and to promote conservation. Resident sharks have been accustomed to human divers, yes, but the result of thousands of divers being introduced to an unforgettable sharky encounter was the creation of two shark marine reserves. These reserves involve the local Fijian people in their protection. Fiji's shark-diving industry is now tackling shark finning, and they have the statistics, research, and dollar figures to back up their argument that sharks are more valuable alive than dead. ❍

Travel Tip:

Beqa is diveable year-round, although conditions vary slightly. Dives are offered on Mondays, Wednesdays, and Fridays. Most operators insist divers have a certain number of logged dives for this two-tank dive, or hire a private guide to accompany them. The safety record for the area is good, but these are wild creatures—dive with your wits about you.

INDONESIA

LIBERTY WRECK

One of the world's best—and most accessible—wreck dives

AVERAGE WATER TEMP: 75°F (23.9°C) **AVERAGE VISIBILITY:** 30 to 100 feet (9 to 30 m)
AVERAGE DEPTH: 60 to 100 feet (18 to 30 m) **TYPE OF DIVE:** Wreck

The U.S.A.T. *Liberty* is New Jersey tough. This U.S. Army transport was built in the Garden State in 1918 and requisitioned for WWII. After being torpedoed by a Japanese submarine in the Lombok Strait in 1942, it limped to a beach, badly damaged, but still intact. And then in 1963, an erupting volcano heaved it 130 feet (39.6 m) offshore, breaking its back and settling it on a bed of black sand, where it currently lies, submerged—but not defeated—in around 80 feet (24 m) of water.

This popular wreck is buzzing with divers, but at 400 feet (122 m) long with a gross tonnage of 6,211 (5,634.5 metric tons), she's a big girl, making up a big site with plenty of room to explore. Divers report this spot is worth several dives before you get the feel for the place.

One reason the *Liberty* is such a popular dive is its accessibility: Beginner divers, intermediate divers, free divers, and night divers explore with equal enthusiasm here.

Another is the fact that she's resting in Bali's Tulamben Bay, which local villages designated a marine protected area in 1979. Rich with plentiful marine life, gorging on food delivered by currents, reef fish (groupers, angelfish, and surgeonfish), corals (sea fans and sponges), and larger fish (sunfish, reef sharks, barracuda) swarm the wreck, a colorful and bountiful underwater garden fitting for a ship that showed heart.

The best time to dive the *Liberty* is October and November, when you can find yourself safe from monsoon season. The monsoons return December through April and mid-July through September, bringing strong winds that make for a rough surface and tricky diving. ❍

What You'll See: Barracuda • Ribbon and Garden Eels • Reef Sharks • Bumphead Parrotfish • Groupers • Nudibranchs • Sunfish • Batfish • Angelfish • Surgeonfish • Trevallies • Sweetlips • Butterflyfish • Pufferfish • Feather Stars • Sea Fans

Coral and algae have encrusted the ship's wheel of the U.S.A.T. *Liberty.*

CANARY ISLANDS

EL BAJON DEL RIO

Mushroom-shaped volcanic rocks shelter a wide variety of marine life.

AVERAGE WATER TEMP: 70°F (21°C) **AVERAGE VISIBILITY:** 32 to 100 feet (9.8 to 30 m)
AVERAGE DEPTH: 53 feet (16.2 m) **TYPE OF DIVE:** Open water

Rising from a golden sand seafloor are several mushroom-shaped volcanic rocks, some 33 feet (10 m) tall. The upper parts of these unusually shaped formations are covered in green and brown algae, as well as bright sponges. Descending farther, divers discover a flurry of marine activity, flashes of silver lit up by the bright sunlight of the Canary Islands. This penetrating underwater light, combined with the odd structures and swirling fish (bream, parrotfish, barracuda, triggerfish), make El Bajon del Rio a popular spot.

The Canary Islands, located off the Moroccan coast in the Atlantic Ocean, have more than 870 miles (1,400 km) of coastline, bathed in a Gulf Stream that brings in a mix of warm water (70°F/21°C in summer) and cooler water (60°F/15.6°C in winter) to its remarkable rocky seascapes.

Tucked into the middle of a strait separating Isla de Lobos and the island of Fuerteventura, El Bajon del Rio benefits from the Fuerteventura's UNESCO Biosphere Reserve status. (Its western coastline—all 62 miles/99.8 km of it—is completely free from human intervention.)

"Divers make up a community, and it's a diverse community. We're photographers. Explorers. Hunters. Some like tropical waters, others prefer the cold. Some people want to go muck diving with little critters, while others want to see big animals. All that diversity we see on land exists under the water, maybe even more so."

—BRIAN SKERRY

A diver swims over volcanic rock formations.

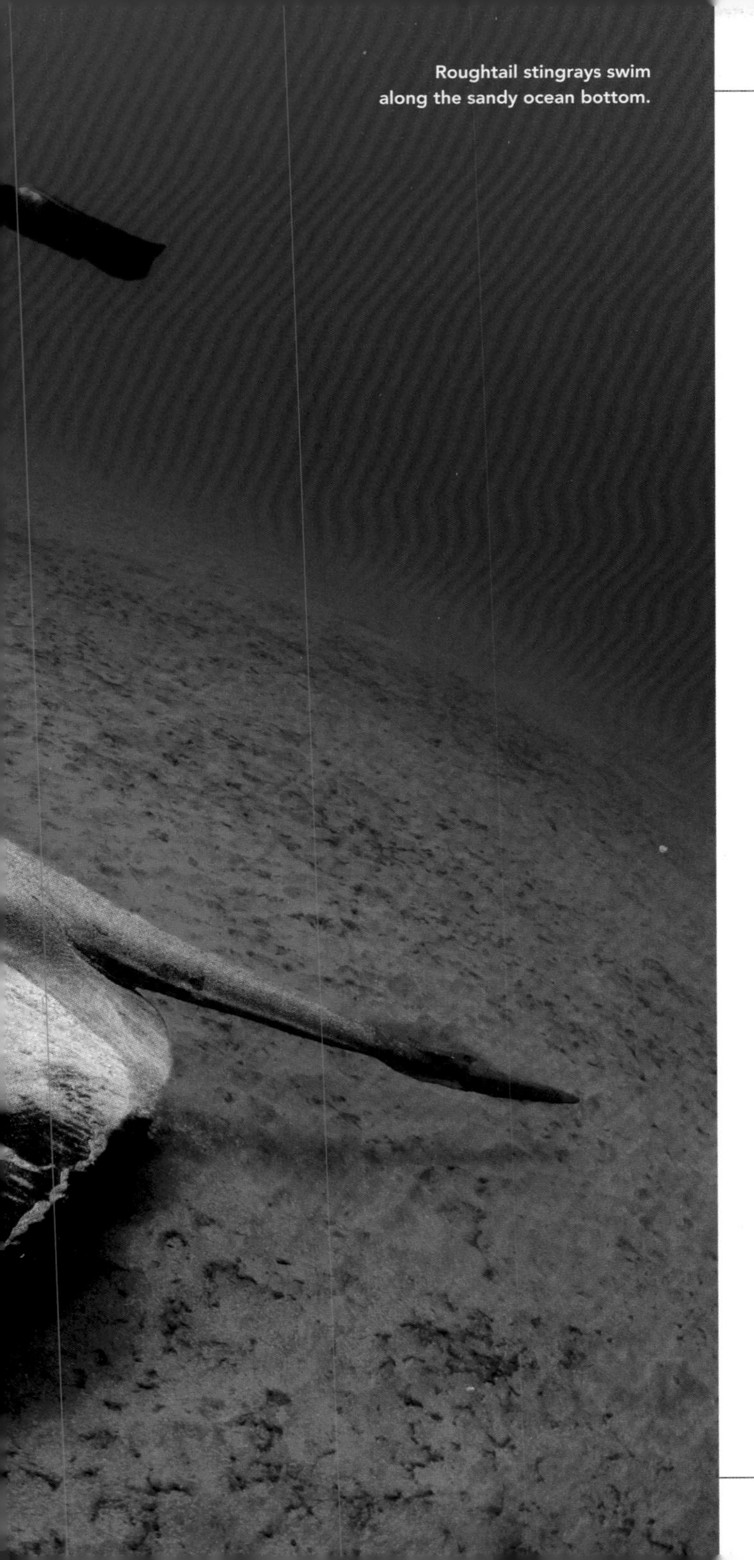

Roughtail stingrays swim along the sandy ocean bottom.

What You'll See: Trumpetfish • Saddled, Blacktail, Zebra, and Cow Bream • Ornate Wrasses • Barracuda • Parrotfish • Jacks • Scorpionfish • Moray Eels • Sponges • Algae • Damselfish • Canarian Lobster • Groupers • Eagle Rays • Stingrays • Triggerfish

Underwater, an exceptional diversity of marine life shelters in the volcanic rocks. Moray eels occupy a lot of crannies and caverns, watching shoals of bream (zebra, cow, and blacktail), and electric-colored parrotfish, ornate wrasses, and scrawled filefish cruise past. Canarian lobsters and stingrays prefer the sand carpet. Combined with the sponge and algae life on the top of the rock formations, there is something to see from every angle.

El Bajon del Rio is an easy dive—when the conditions are right. Currents and tides, particularly in winter, can make it more challenging, so utilize local knowledge and dive to your limits. The best time to dive here is June through October, when water temperatures are at their warmest and currents are mild. And getting here is simple: There is no shortage of flights to Fuerteventura, and from there it's a five-minute boat ride from the Port of Corralejo to El Bajon del Rio. ❍

Travel Tip:

Eight airports serve the Canary Islands area, so getting here is easy. Though diving is best June through October, airfare and hotel prices are at their lowest in November and early December, as well as April through June. Carnival celebrations take place in February and March.

CAYMAN ISLANDS

BLOODY BAY WALL

The Caribbean's most breathtaking drop-off

AVERAGE WATER TEMP: 80°F (26.7°C) **AVERAGE VISIBILITY:** 200 feet (61 m)
AVERAGE DEPTH: 25 feet (7.6 m) **TYPE OF DIVE:** Wall

Rumored to be named for bloody pirate battles that took place in the area, Bloody Bay Wall is ideal for beginner divers due to its shallow entrance, calm waters, and vivid marine life. This undersea spacewalk begins at a shallow depth of 20 feet (6 m), running parallel to the shoreline before plunging dramatically into a clear blue abyss more than 1,000 feet (305 m) deep. (Some estimates suggest it's between 3,000 to 6,000 feet/914 to 1,829 m deep.) Wherever the bottom of Little Cayman's Bloody Bay Wall actually lies, divers report feeling as though they're hanging in space, suspended, transported, with 200 feet (61 m) of visibility above, to either side, and below.

Once you've recovered from the breathtaking shock of the vertical cliff face, look around—you won't be disappointed. Far from a sheer rock wall, Bloody Bay teems with life. Sponges seem to glow in gemstone colors of citrine, emerald, ruby, and sapphire. Fan coral stretches out delicate branches, while Nassau groupers, green turtles, and triggerfish cruise alongside you.

Part of the Bloody Bay Marine Reserve, the wall and its nearby maze of coral canyons—a string of more than a dozen dive sites that run along its edge—offer a variety of cruisy dive sites with consistently clear, warm water and plenty to see.

You can dive the Bloody Bay Wall—and other sites around the Cayman Islands, just 480 miles (772.5 km) south of Miami and 170 miles (273.6 km) northeast of Jamaica—all year round. But keep in mind that the islands are situated in the hurricane belt and weather can be variable from June to November. Access to Little Cayman is by air or liveaboard from Grand Cayman. ❍

What You'll See: Fan Coral • Tube Sponges • Triggerfish • Green Turtles • Nassau Groupers • Horse-Eye Jacks • Damselfish • Eagle Rays • Angelfish

A hawksbill turtle explores the water in the shadow of a dive boat.

PHILIPPINES

TUBBATAHA REEF NATURAL PARK

A wide variety of sites and species make this an ideal liveaboard for beginners.

AVERAGE WATER TEMP: 83°F (28.3°C) **AVERAGE VISIBILITY:** 130 feet (39.6 m)
AVERAGE DEPTH: 50 feet (15 m) **TYPE OF DIVE:** Open water

Tucked away in a remote corner of the Philippines, Tubbataha Reef Natural Park is often overlooked by divers. It shouldn't be. This 247,100-acre (1,000 sq km) UNESCO World Heritage site is home to a rich diversity of marine life—both large and small—including 600 species of fish (Moorish idols, groupers, and Napoleon wrasses, among many), 360 species of coral (including 90 percent of all the coral species found in the Philippines, like sea fans and barrel sponges), and 11 species of shark (zebra, hammerhead, whitetip and blacktip reefs), as well as hawksbill and green turtles. Whale sharks and tiger sharks make occasional appearances, and manta rays fly through to visit cleaning stations.

Located about 90 miles (145 km) southeast of Palawan in the center of the Sulu Sea on a line of extinct underwater volcanoes, this is one of the Philippines' oldest ecosystems. The name Tubbataha comes from the Sama people, roughly translating to "long reef exposed at low tide." Two coral atolls (North and South Tubbataha) are split by a five-mile-wide (8 km) channel where schools of fish cloud crystal visibility extending more than 100 feet (30 m).

This is a good liveaboard spot for beginners: The large variety of sites (atolls, reefs, walls, shallow lagoon cruises, and pinnacles) paired with a high density of species makes for a mind-blowing visual spectacle. There is always something to see. Dive to your limits, but when you're ready to try your hand at a little current, you'll be rewarded with visits from some of the bigger species.

Tubbataha is home to 90 percent of the coral species native to the Philippines.

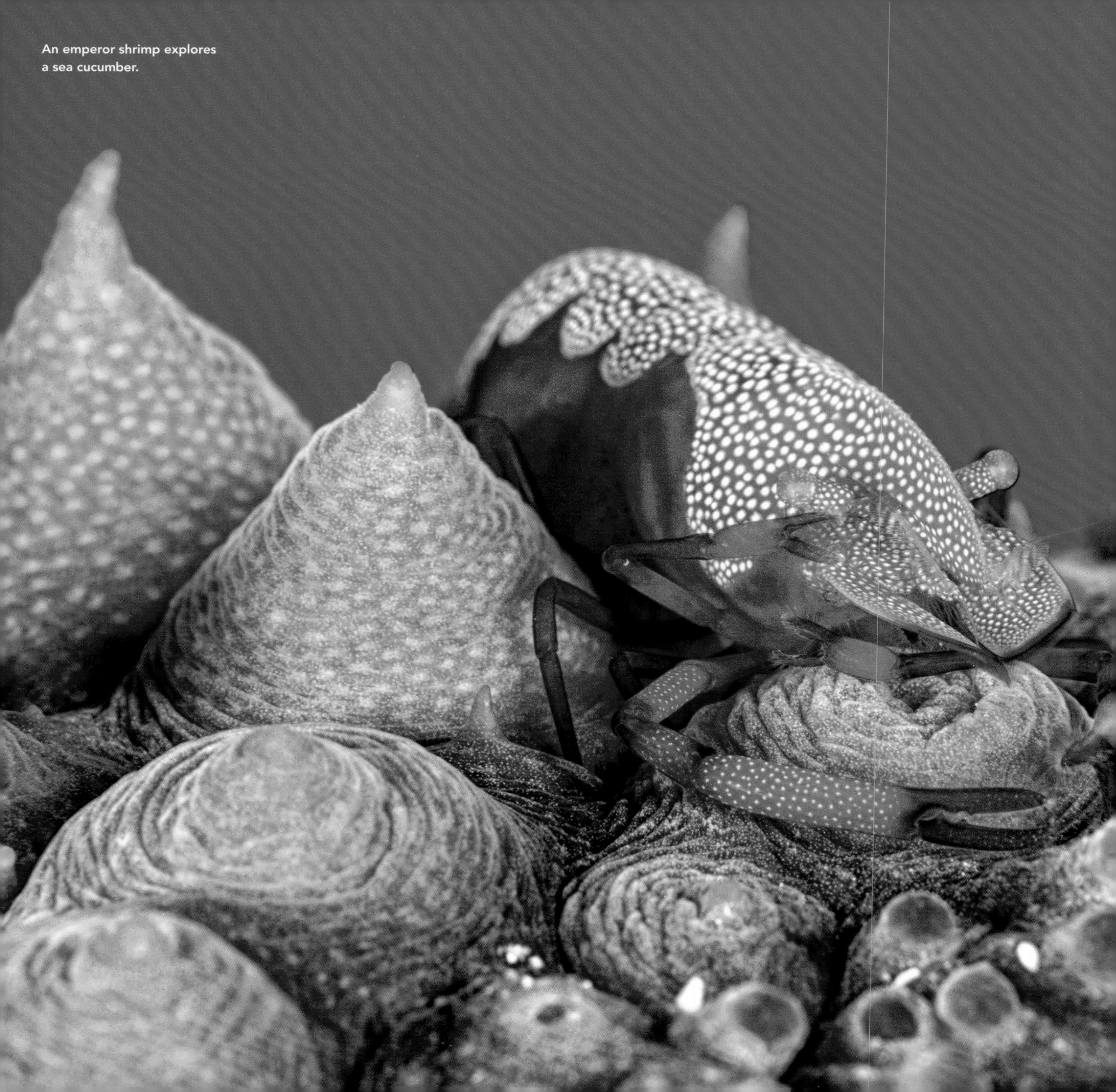

An emperor shrimp explores a sea cucumber.

What You'll See: Whale Sharks • Manta Rays • Moorish Idols • Jacks • Snapper • Barrel Sponges • Sea Fans • Barracuda • Tuna • Eagle Rays • Groupers • Hammerheads • Whitetip Reef Sharks • Blacktip Reef Sharks • Gray Reef Sharks • Napoleon Wrasses • Tiger Sharks • Trevallies • Zebra Sharks • Sweetlips • Bumphead Parrotfish • Green and Hawksbill Turtles

The oblong-shaped North Atoll is one of the most diverse coral reef systems in the Philippines, a mix of coral gardens, sandy lagoons, sea grass beds, and steep walls.

Bird Islet, a wall with crevices and overhangs, features hard corals, garden eels, giant reef rays, and sharks. Amos Rock, with its stiffer current, also boasts its fair share of sharks, including gray and whitetip reef sharks, as well as brilliant coral (whip and sea fans), Napoleon wrasses, snapper, groupers, and mackerels. This is a favorite spot for night divers.

The triangular-shaped South Atoll is a rich reef, home to nooks and overhangs favored by the likes of lobster and whitetip reef sharks. Lighthouse Islet is a gentle dive with a shallow, sloping reef filled with feeding turtles, stingrays, angelfish, and anemones, while T Wreck—a wall covered in barrel sponges, sea fans, and coral—attracts larger pelagics, like reef sharks, groupers, and triggerfish. ❍

Travel Tip:

Liveaboards are the only way to dive Tubbataha Reef. Trips typically last from six to 13 nights, running between mid-March to mid-June. April is usually the best time to dive due to calm conditions. The rest of the year, the area is closed to divers to better protect and conserve it.

CUBA

LOS JARDINES DE LA REINA

The Caribbean the way it used to be

AVERAGE WATER TEMP: 82°F (27.8°C) **AVERAGE VISIBILITY:** 100 feet (30 m)
AVERAGE DEPTH: 80 feet (24 m) **TYPE OF DIVE:** Liveaboard

Ask anyone who grew up diving in the Caribbean and they'll tell you the same thing—it's not the way it used to be. Ask any diver who's been lucky enough to explore Cuba's Los Jardines de la Reina (the Gardens of the Queen), and they'll say the same thing: *This* is the way the Caribbean used to be. And the way it should be.

Located a mere 100 miles (160.9 km) from the United States, and 50 miles (80.5 km) off Cuba's southern coastline, this 840-square-mile (2,175.6 sq km) marine reserve (encompassing mangroves, reefs, and islands) is the largest and most fiercely protected marine park in the Caribbean.

Fidel Castro recognized the importance of protecting this area back in 1996, and the United States might want to take notice that 25 percent of Cuba's marine waters are protected, compared with less than 2 percent of waters along the continental United States. That effort has reaped dividends. (It's also worth noting that Castro's love for diving almost killed him. Political rumor has it that the CIA once tried to assassinate him by putting tuberculosis in his regulator.)

Fragile and threatened elkhorn coral, the underwater canary in the coal mine, is virtually absent from most of the Caribbean, but it's not only thriving in the Gardens of the Queen, it's increasing, and the same goes for shark populations (especially silky sharks), the size of groupers in the area (black and Nassau), and reef ecosystems.

Named by Christopher Columbus for his patron, Queen Isabella, on his second trip to the region in 1494, getting to the Gardens still involves a modicum of difficulty.

A gap-jawed Nassau grouper swims above an array of coral.

Caribbean reef sharks patrol the national marine park's waters.

What You'll See: Silky, Nurse, Caribbean Reef, and Hammerhead Sharks • Saltwater Crocodiles • Jacks • Nassau, Goliath, Black, and Tiger Groupers • Eagle Rays • Sea Fans • Azure Vase Sponges • Elkhorn Coral • Tarpon • Hawksbill, Loggerhead, and Green Turtles • Caribbean Reef Squid • Queen Conchs • Blue-striped Grunts • Porkfish

U.S. and Cuban relations, for example, seem to have a state of perpetual motion. The dive numbers are restricted and the only diving available is by liveaboard. However, for the lucky few divers who are able to make the trip, the sheer quantity and variety of marine life seen on the three to four daily dives in the gardens is staggering.

There are approximately 120 dive sites at Los Jardines, which are regularly rotated to provide respite to each area. The majority of dives can be broken into two themes: sharks and reefs.

Black Coral I and II, Pipin, and Five Seas are a few of the best sharky spots. A variety of sharks are ever present, but the silky sharks steal the show. Here divers can join in a spinning carousel of up to 30 silky sharks, watched by enormous groupers (Nassau, goliath, black, and tiger) whose fearlessness and size are a testament to the fact that they aren't being hunted (at least by humans).

Travel Tip:

The Gardens are a year-round destination. It's best December through April, with May through October being the rainy season. Hurricanes occasionally come through in October, as well.

A shrimp perches on its host anemone.

A once-in-a-lifetime experience: diving with saltwater crocs

For reef dives, Farallon and Vicente are favorites, with rich gardens of gorgonians, azure vase sponges, elkhorn, and staghorn corals. And keep an eye out for the shy saltwater crocodiles that call this area home—they're very likely watching you.

With little current, and protection from waves and rough seas, Los Jardines de la Reina is an ideal spot for beginner divers to dive, dive, and dive some more. ❍

"There were so many sharks in Gardens of the Queen I couldn't see them all."

—ERIKA BERGMAN, SUBMARINE PILOT AND NATIONAL GEOGRAPHIC EXPLORER

HAWAII

FIRST CATHEDRALS

Dive through tunnels created by molten lava flowing into the sea.

AVERAGE WATER TEMP: 75°F (23.9°C) **AVERAGE VISIBILITY:** 90 feet (27 m)
AVERAGE DEPTH: 25 to 60 feet (7.6 to 18 m) **TYPE OF DIVE:** Cavern

Descending down First Cathedrals, sunlight filters through broken lava, casting eerily poignant shadows reminiscent of stained-glass windows, hence its name. Like all Hawaiian Islands, Lanai was formed by lava, as was its underwater reefs—twisted formations frozen when blazing hot lava surged into the sea. The cooled rock created sanctuaries for marine species, both the small (like the gold lace nudibranch) and the large, like the "snowbirding" humpback whales that flock to Hawaii's warm, deep waters during the winter months. It also created an underwater wonder for divers. If you want to see new or rare fish, this is the place. The island is known for its excellent and reliable visibility, due to minimal runoff, which is filtered by the volcanic cliffs.

In the protected waters off Lanai, near the adjacent and popular bays of Hulopo'e and Mānele, lies its premier dive spot: the towering dome of First Cathedrals (also known as First Cathedral and Cathedral I). This 50-foot-deep (15 m) cavern has a soaring, 25-foot-high (7.6 m) ceiling, beckoning divers looking to explore the Earth turned inside out. Far from a smooth lava tube, First Cathedrals is full of passageways, ridges, and tunnels, not to mention the aptly named Shotgun exit, which needs to be timed with the surge.

Beginner divers can explore the entrance of the Cathedral, marveling at the sunlight streaming through lava windows, illuminating the cavern with a cinematic quality. Whitetip reef sharks rest on the bottom, tucked under rock ledges shared by lobster and crabs. Outside of the grotto, reef fish (like butterflyfish and angelfish), as well as the occasional turtle or dolphin, delight divers. ❍

What You'll See: Green Turtles • Gold Lace Nudibranchs • Whitetip Reef Sharks • Butterflyfish • Saddleback Butterflyfish • Spinner Dolphins • Lobsters • Banded Angelfish • Humpback Whales (in winter months)

A diver explores one of the First Cathedrals' many passageways.

REPUBLIC OF KIRIBATI

KIRITIMATI

One of the last untouched reefs in the world

AVERAGE WATER TEMP: 82°F (27.8°C) **AVERAGE VISIBILITY:** 125 feet (38 m)
AVERAGE DEPTH: 30 to 130-plus feet (9 to 39.6+ m) **TYPE OF DIVE:** Shore, boat, and reef

Most people have never heard of Kiritimati, even though it's the largest coral atoll in the world. This 150-square-mile (388.5 sq km) speck in the Pacific Ocean (with a lagoon that's nearly the same size) is part of the Republic of Kiribati, a windswept archipelago located roughly 1,300 miles (2,092 km) south of Hawaii and northeast of Fiji, the two countries that (at the time of writing) offer weekly flights to Kiritimati.

Kiribati, part of the Northern Line Islands, is home to one of the South Pacific's largest marine reserves, covering nearly 158,000 square miles (409,218 sq km), and host to more than 500 species of fish and 20 species of seabirds. Kiribati's string of 32 atolls and one coral island straddles the Equator.

Kiritimati is a respelling of the word "Christmas" in the Kiribati language, hence the nickname "Christmas Island," which is often confused with Australia's Christmas Island (page 214). It was named by British explorer Captain James Cook, who stopped by Kiritimati on Christmas Eve 1777.

Kiritimati is all about the sea—it is not only surrounded by it but also rarely out of sight or hearing distance of the ocean. Studded by tidal flats, salt pans, and ponds, Kiritimati looks like a mottled turtle shell from the air, the entire island a wildlife sanctuary with white and golden beaches, crystal lagoons, and surrounded by coral gardens plunging into the open ocean.

Given the large lagoon, Kiritimati is ideal for beginner divers—the entire perimeter of the island is one large dive site, frequently referred to as the last untouched reef in the world. Divers access this playground by shore, large outrigger canoes, or dive boats, and descend into a lush garden of marine life. Larger species include green turtles (which nest on the island), spinner dolphins, manta rays, octopuses, groupers, spotted eagle

An aerial view of Kiritimati's unique topography

What You'll See: Ghost Crabs • Coconut Crabs • Strawberry Land Hermit Crabs • Barracuda • Moorish Idols • Trevallies • Manta Rays • Spinner Dolphins • Spotted Eagle Rays • Nudibranchs • Surgeonfish • Green Turtles • Bonefish • Whale Sharks • Plate and *Acropora* Coral • Butterflyfish • Anemones

rays, dragon moray eels, bonefish, barracuda, snapper, and seasonal whale sharks.

With all that on offer, you still don't want to overlook the smaller critters. Unspoiled *Acropora* and plate corals provide homes to butterflyfish, gobies, anemones, nudibranchs, Moorish idols, and flame angelfish. The island is also home to a number of crustaceans, including strawberry land hermit crabs, ghost crabs, and coconut crabs.

This beautiful paradise also has a nuclear test history (from both the United Kingdom and the United States in the 1950s and 1960s), but its greatest threat is yet before it: Rising sea levels are an immediate menace to the low-lying Kiritimati and the Republic of Kiribati, its coral atolls in danger of being reclaimed by the ocean. Kiritimati's unique topography, as well as its isolation—the two things that shaped its character—are now working against it, and the clock is ticking. ❍

Travel Tip:

Kiritimati is a year-round destination, with the water temperature mirroring the air temperature and almost daily showers. Minke whales and dolphins are frequent visitors December through March (which is also the typhoon season), and November through April is the best time to see whale sharks.

A bohar snapper swims above thriving coral reefs.

PAPUA NEW GUINEA

KIMBE BAY

Rich, diverse marine life and a WWII aircraft wreck

AVERAGE WATER TEMP: 84°F (28.9°C) **AVERAGE VISIBILITY:** 95 feet (30 m)
AVERAGE DEPTH: 80 feet (24 m) **TYPE OF DIVE:** Open water, shore, and wreck

Kimbe Bay, a marine protected area, holds nearly 60 percent of the coral species found in the Indo-Pacific, and more than half of the world's coral species. It's rich, remote, and worth the journey to swim past forests of red sea fans, pink sea whips, staghorn coral, and large coral bommies in pristine condition.

Located on the north coast of New Britain, east of Papua New Guinea and its largest island, Kimbe Bay's calm, shallow coral gardens have a legendary reputation among divers. Accessible by shore, boat, or liveaboard year-round, there is something on offer for every type of diver.

Susan's Reef is a favorite and a well-photographed site. Two seamounts connected by a saddle are home to anemones, sea fans, and colorful reef fish.

South Emma, another popular dive spot, is a seamount with bigger action (barracuda, reef sharks, trevallies, and jacks often put in appearances), adding to the color of soft coral and barrel sponges. Plus, keep an eye out for the swim-through cave.

One of Kimbe Bay's most famous spots, however, is Zero Wreck, a submerged Japanese World War II fighter plane. Visibility can vary at this site, due to its proximity to shore, but this intact wreck lies upright on a sandy bottom in the shallows (56 feet/17 m). Research indicates this wreck was the result of a (very) controlled landing, so the plane is in remarkable condition, with little coral growth over the plane's exterior or three-bladed propeller. Look for the large anemone that lives behind the pilot's seat in the open cockpit as you explore the craft. ❍

What You'll See: Staghorn Coral • Sea Fans • Sea Whips • Barrel Sponges • Reef Sharks • Orcas • Jacks • Barracuda • Tuna • Seahorses • Trevallies • Fusiliers

A Japanese World War II fighter plane rests in its watery grave.

A diver moves in close for a look at a resting green turtle.

BORNEO

PULAU SIPADAN

All the marine life you've ever dreamed of seeing—on one island

AVERAGE WATER TEMP: 82°F (27.8°C) **AVERAGE VISIBILITY:** 66 feet (20 m)
AVERAGE DEPTH: 75 feet (22.9 m) **TYPE OF DIVE:** Open water

Would the diving world have discovered Pulau Sipadan (Sipadan Island) without Jacques Cousteau? Perhaps, but it would have taken a while to plumb the underwater secrets of this unassuming, tiny, Tic Tac–shaped coral island without him.

Tucked an hour's boat ride off the northeastern corner of Borneo in the Celebes Sea, you can walk the entire perimeter of Sipadan in less than 30 minutes. Most visitors turn their eyes upward on this declared bird sanctuary (established in 1933), searching out kingfishers, wood pigeons, sea eagles, sunbirds, and starlings that make Sipadan's dense vegetation home.

But the hidden gems are encased by the island's crystal clear waters, a rich underwater world Cousteau referred to as an untouched piece of art, thereby registering Sipadan on every diver's radar.

What You'll See: Barracuda • Bumphead Parrotfish • Jacks • Turtles (Green and Hawksbill) • Frogfish • Leaf Fish • Pygmy Seahorses • Sharks (Zebra, Whitetip Reef, Gray Reef, Hammerhead) • Manta Rays • Nudibranchs • Coral (Sea Fans, Staghorn, Table, Barrel Sponges)

A school of green bumphead parrotfish, the largest of its species

And UNESCO's, which named the area a World Heritage site in 2005, a year after all the island's resorts and facilities were shut down to preserve its wealth of lush landscape and diverse wildlife living on and off the islands.

Below the surface, things get interesting, with nearly every marine species you've ever dreamed of vying for your attention. Sipadan safeguards more than 3,000 species of fish and hundreds of species of coral, and has antipoaching groups permanently stationed on the islands to watch over the marine reserve's treasures. In a nutshell, the island's waterways have everything you've ever wanted to see under the sea. Deep walls dropping more than 1,900 feet (579 m) pull in the pelagics, while the colorful and shallow coral gardens have a habit of stretching out safety stops.

The Sipadan Barrier Reef is the largest in Southeast Asia, with a stunning array of biodiversity. With many dive sites to choose from, you'll never run out of things to see, from manta ray cleaning stations to devil rays and tons of schooling sharks. And some of the world's best macro life lives here as well: pygmy seahorses, flamboyant cuttlefish, and mandarinfish, just to name a few.

The Coral Garden is a healthy, shallow reef perched on an underwater wall that's ever present, but unnecessary to descend to get the best from this location. Divers can drop to a comfortable 16 to 33 feet (4.9 to 10.1 m) to observe strong coral growth, as well as the life it supports, including feather stars, angelfish, unicorn fish, butterflyfish, triggerfish, and humphead wrasses. Green turtles are ever present here, as they are everywhere in Sipadan. The island is a breeding ground for the species, who lay eggs on the beaches between August and September (beaches have restricted access to protect their nesting sites).

The Turtle Patch is a similarly turtle-plush area and another wall-that-doesn't-have-to-be-a-wall-dive spot. Located on the southeast corner of the island, and best dived in the mornings, divers can float over shallow coral gardens on top of a wall, letting the current

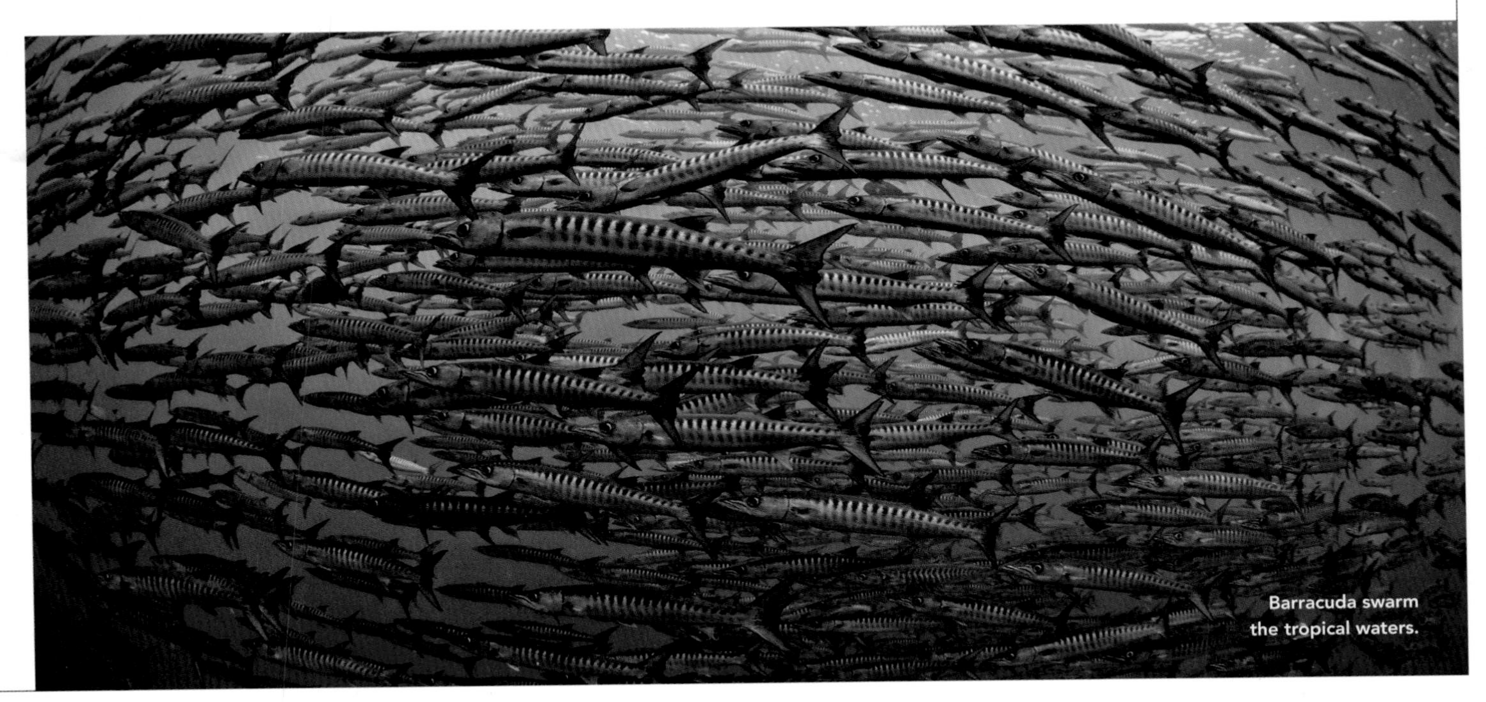

Barracuda swarm the tropical waters.

Travel Tip:

Sipadan is diveable year-round, July and August being the best months. April through September, green and hawksbill turtles come ashore to nest. January and February can be unsettled and rainy. Avoid (or book well in advance for) March to August, Christmas, the New Year, and the Chinese New Year periods.

drift them lazily along the shallows, observing staghorn, table, and leather corals. Turtles feed on the coral or rest in sandy patches, oblivious to passing divers, although the orange spinecheek anemonefish will keep a close eye on you from their colorful homes.

Barracuda Point kicks things up a level. This spot is home to schools—nay, tornadoes—of its namesake fish (chevron and blacktail), as well as jacks, snapper, red-toothed triggerfish, and gray reef sharks. This is a northern wall site, about 33 feet deep (10 m) along the top of the wall, but don't descend too deeply unless you want to battle with currents that can get downright vicious.

At the Drop Off, the signature dive of Sipadan, you'll find a mesmerizing underwater world surrounding a 1,900-foot (579 m) wall. The wall itself is covered in coral and sponges, and offers the ultimate in night diving where you can look into nooks and crannies for nocturnal crab and shrimp. In daylight, hundreds of jacks, trevallies, or barracuda circle around you—often distracting you from the whitetip sharks, gray reef sharks, and green turtles lying in wait.

For a dive of another color, visit Whitetip Avenue, where a guide will determine your route to find bigeye trevallies and large bumphead parrotfish. Even beginner divers are encouraged to explore the deep side of the drop-off here, where ledges, chimneys, and other various landscapes host black coral, sea fans, groupers, emperor angelfish, boxfish, and scorpionfish, among others.

Accessed by liveaboard or a day visit from a neighboring island (Mabul, Kapalai, or Mataking), the trick with Sipadan is the permit system, which allows only 120 divers a day, with restrictions on depth (131 feet/40 m) and requirements around dive gloves. Booking in advance is essential here. But the longer you stay, the greater your opportunity to dive this head turner of a destination. ○

A small necklace sea star sits atop a larger blue sea star.

STATIA

CHARLIE BROWN WRECK

One of the biggest—and least visited—wrecks in the Caribbean

AVERAGE WATER TEMP: 81°F (27°C) **AVERAGE VISIBILITY:** 100 feet (30 m)
AVERAGE DEPTH: 50 to 100 feet (15 to 30 m) **TYPE OF DIVE:** Wreck

The Caribbean's popularity is also its curse. Many divers feel there's nothing left to be explored, but there is always something new to see under the sea.

Take Statia, for example. If you're uncertain where that is, you're not alone. St. Eustatius (the official name for this special municipality of the Netherlands) is located south of Anguilla and north of St. Kitts and Nevis.

It's said the only thing you won't see on Statia are crowds—above water or below. This tiny, one-town (Oranjestad) island has a population of around 4,000 people, and a larger-than-life, nearly perfectly shaped 2,000-foot (609.6 m) volcano called The Quill that looms over the landscape.

Statia is another Christopher Columbus "discovery," and after his sighting in 1493 the island was tossed around like a rugby ball between the French, British, and Dutch, changing hands more than 22 times before the Dutch nabbed possession in 1756.

Once a bustling port, Statia now gets more visits from nesting hawksbill, leatherback, and green turtles than from merchant ships. Protected by the St. Eustatius National Marine Park, a 63,011-acre (255 sq km) sanctuary established in 1996, marine life has been allowed to flourish, and almost all of Statia's 36 dive sites are moorings to protect the reef from anchorage.

But—surprisingly—Statia's well-protected reefs and abundant marine life aren't the only feathers in its oceanic cap. Divers come here especially to see the C.S. *Charles L. Brown* (affectionately referred to by locals as the *Charlie Brown*). The *Charlie Brown* operated as a cable repair ship until 2002, when Statia purchased the craft for a dollar,

One of the *Charlie Brown*'s
anchors on the seafloor

The shipwreck's massive hull offers plenty for divers to explore.

What You'll See: Wrecks • Hawksbill, Green, and Leatherback Turtles • Hermit Crabs • Land Crabs • Barracuda • Jacks • Sea Fans • Sponges • Reef Sharks

then purposefully sank it a year later to serve as a living reef on the ocean floor.

This 327-foot (99.7 m) wreck is impressive and intact. Jacks and barracuda school above the ship, with the occasional reef shark patrolling nearby. Sea fans and sponges are growing from the deck, and turtle sightings are common.

The *Charlie Brown* is lying on its side on a sandy patch 50 to 100 feet (15 to 30 m) deep. Novice divers can enjoy circling above the wreck, while more experienced divers can descend to depth and discover what awaits within its hull. (The interior is in good shape and fairly easy to explore. As always, exercise caution and dive to your limits.) Keep an eye out for the smaller stuff that sticks close to the sea fans and sponges. And look for any flashes of blue. Statia is famous for its blue glass beads, which were used as local currency in the 17th century. Storms stir up the sand, often revealing these treasures (the only artifacts you can remove from Statia). Legend has it that you don't find blue beads—they find you, and if they do, you will always return to Statia. ❍

Travel Tip:

Statia is diveable all year long, but hurricane season is June through November, and more rain tends to fall October to January. Divers pay a $6 fee per dive to visit the marine park, and diving without a guide is not allowed.

NEW ZEALAND

POOR KNIGHTS ISLANDS

A rich marine reserve with dramatic underwater formations

AVERAGE WATER TEMP: 65°F (18.3°C) **AVERAGE VISIBILITY:** 65 to 100 feet (19.8 to 30 m)
AVERAGE DEPTH: 15 to 130 feet (4.6 to 39.6 m) **TYPE OF DIVE:** Open water

New Zealand's Poor Knights Islands are dramatic—not in the overused adjective sense of the word, but in its truest meaning: This place is filled with action. From its rich and anguished Maori history (resulting in the islands being *tapu*, or sacred, forbidden), to its violent volcanic origins producing cliffs, caves, and undersea arches, to a crush of marine life, it's little wonder Jacques Cousteau named the Poor Knights one of the top 10 dive sites in the world.

Located 14 miles (22.5 km) off the North Island's east coast, the Poor Knights are the remnants of 11-million-year-old volcanoes carved into tunnels, caves, and cliffs that provide a home to a wide variety of species brought together by the mix of tropical and temperate waters.

Here, orcas hunt squadrons of stingrays past gorgonian fields, while more than 125 species of fish (including grouper, sunfish, trevally, and snapper) cruise through kelp forests or over sponge gardens. Electric blue maomao crowd under their namesake arch, 65 feet (19.8 m) across with shafts of sunlight piercing through holes in the roof. Another impressive formation, the Rikoriko Cave, is big enough to sail a ship into. This sea cave, the largest surveyed in the world, is 262 feet (79.9 m) wide and 427 feet (130 m) long.

Because of the rich marine life, the Poor Knights were made a marine reserve in 1981, and the islands themselves were designated a nature reserve. ❍

What You'll See: Blue Maomao • Snapper • Trevallies • Bronze Whaler Sharks • Manta Rays • Bull Rays • Moray Eels • Sea Fans • Black Coral • Orcas • Pilot Whales • Nudibranchs • Triplefin Blennies • Striped Boarfish • Gold-Ribbon Groupers • Sunfish

A nudibranch (mollusk) guards its eggs.

MALAYSIA

SWALLOW REEF

The jewel of the South China Sea

AVERAGE WATER TEMP: 80°F (26.7°C) **AVERAGE VISIBILITY:** 65 to 130 feet (19.8 to 39.6 m)
AVERAGE DEPTH: 60 to 130 feet (18 to 39.6 m) **TYPE OF DIVE:** Open water and reef

Layang Layang ("Place of Swallows" in Malay, also known as Swallow Reef) is another military installation, a (nearly) man-made atoll of 13 linking coral reefs that is home to an airstrip, Malaysian base, and dive resort. This five-square-mile (13 sq km) sliver of land is part of the South China Sea's Spratly Islands, a hotly contested piece of real estate 186 miles (299 km) off the Borneo coast. Luckily, fish don't give a fin about politics, and they flock to the rich, pristine waters in droves, protected in part by the resident military base.

Swallow Reef is surrounded by a variety of dive options, ranging from calm, shallow lagoons on one side of the island to the plunging 6,500 feet (1,981 m) of open ocean on the other, a deep drop-off that attracts large pelagics.

Scalloped hammerheads are frequent visitors, as well as trevallies, zebra sharks, and green and hawksbill turtles. Sea fans stretching 10 feet (3 m) across shelter smaller creatures like nudibranchs, seahorses, pipefish, and anemone shrimp.

Flourishing coral gardens in 15 to 50 feet (4.6 to 15.2 m) of bath-warm water make for a tranquil cruise, spying corals like elkhorn, warty, bottlebrush, and sea whips along the way.

Feeling lucky? Keep a weather eye out for the island's occasional visitors: thresher sharks, whale sharks, and melon-headed whales.

Layang Layang's dive resort is open from March to October, and usually closes November to February due to monsoon rains. The resort is open during the hammerheads' favorite time to visit—March through May. As this is a military base with a solitary dive resort, make travel plans well in advance. ❍

What You'll See: Hammerheads • Zebra Sharks • Gray Reef Sharks • Silvertip Sharks • Manta Rays • Marbled Rays • Eagle Rays • Pygmy Devil Rays • Cuttlefish • Seahorses • Pipefish • Nudibranchs • Sea Fans • Sea Whips • Elkhorn Coral

Rich and colorful coral reef walls offer plenty of photo opportunities.

INDONESIA

WAKATOBI NATIONAL PARK

A huge and historically protected area that's hard to beat for coral and tropical fish

AVERAGE WATER TEMP: 82°F (27.8°C) **AVERAGE VISIBILITY:** 49 to 260 feet (14.9 to 79 m)
AVERAGE DEPTH: 15 to 130 feet (4.6 to 39.6 m) **TYPE OF DIVE:** Open water and shore

Wakatobi has had a long time to flourish. Established as a national park in 1996, this 3.4-million-acre (13,759 sq km) area has protected reefs and limited fishing, backed by the local communities who share in the dive tourism revenue. Combine that with the fact that Wakatobi isn't easy to reach, and you have an intoxicating cocktail for divers: pristine, healthy, and uncrowded reefs.

Wakatobi (an amalgamation of the names of the four main islands in the Tukang Besi archipelago—Wangi-wangi, Kaledupa, Tomia, and Binongko) was one of Jacques Cousteau's favorite places to dive—some reports say it was *the* favorite. Located at the southeastern tip of Sulawesi, the world's 11th largest island in the Banda Sea, Wakatobi has had the benefit of being championed by the Indonesian government, the Nature Conservancy, and the World Wide Fund for Nature, among others. The result is a garden of coral with more species than almost any other place on the planet, comprising fringing, atoll, and barrier reefs. And those reefs are unlike any other in the region—they are fossilized reefs, which creates a spectacularly clean diving environment from lack of erosion.

Once you've reached Wakatobi, the rest is an easy walk in the (marine) park, with gentle currents, good visibility, warm water, and plenty to see. The area can be dived via liveaboard; there is also one resort with excellent shore and boat dive options. At Wakatobi, the devil is in the details. Small creatures, tropical fish, coral, banded sea snakes, spinecheek clownfish, sea whips, and gorgonians impress more than large pelagics.

The rather dully named House Reef is anything but. Just 65 feet (19.8 m) from your bed, this has to be the best house reef in the world, beginning with a sandy bottom (keep an

There are three scheduled boat dives a day at the Wakatobi Dive Resort.

Blue sea squirts and yellow cave coral add their rich colors to the park.

What You'll See: Pygmy Seahorses • Frogfish • Spine-Cheek Clownfish • Sea Fans • Sea Whips • Sponges • Batfish • Crocodilefish • Cuttlefish • Banded Sea Snakes • Fusiliers • Butterflyfish • Eagle Rays • Hawksbill Turtles • Squid

eye out for blue-spotted rays), cruising to a 40-foot (12 m) drop-off with a wall full of goodies: brightly colored coral, anemones, scorpionfish, eagle rays, hawksbill turtles, and crabs. This is a favorite first and last dive of the day for divers staying at the resort.

Roma is another popular spot with visiting divers. A center pinnacle seems to gather life to itself including schools of fusiliers, banded sea snakes, and triggerfish. Nearby bommies and reef boulders are also home to a wide variety of marine life, from leaf fish to comet fish.

Cornucopia lives up to its name in color and marine life diversity. The current that passes through here encourages a rich menu of plankton that attracts a large array of residents, including the occasional whitetip and nurse sharks, and rays.

The Coral Garden offers plenty to see after you

Travel Tip:

Wakatobi can be dived all year round, but the best time of year is between March and December. July and August are coral spawning season. Although this attracts a large number of fish, it can also be windy with rougher waters. The dive resort is closed during January and February due to the rainy season. Getting here can be tedious—entry points include Manado, Wangi-wangi, Kendari, or Makassar, with an additional bus or ferry transfer. One work-around is the twice-weekly private air charter from Bali, which trims the journey.

A diver captures underwater photos of the rich coral reefs.

A pygmy seahorse blends into its coral surroundings.

make it into the calm waters from a current-swept sand ledge. Here, cleaner wrasses are hard at work, while garden eels and goatfish idle by.

There's also Table Coral City, where large growths of mushroom-shaped table coral shelter patches of staghorn and cabbage reefs. Along this dive site's slopes, you'll also see sponges, sea fans, and anemones playing host to clownfish. Look up, where you might spot barracuda swimming overhead.

Wakatobi is also home to plenty of black coral, particularly at Black Forest, a spot where outcroppings offer tons of the stuff. This is a shallower dive with plenty to see, including Oriental sweetlips, clown triggerfish, Moorish idols, and sergeant majors.

Closer to the resort you'll find The Zoo, a mecca for macrophotographers. Located in a patch of reef nestled in a sandy bay, this spot is home to frogfish, leaf scorpionfish, and ornate ghost pipefish.

Wakatobi is a diver's dream you won't want to wake from. Plan to spend time here—however many days you book, it won't be enough. ❍

BAHAMAS

RAY OF HOPE WRECK

Wreck diving and a feed with Caribbean reef sharks

AVERAGE WATER TEMP: 80°F (26.7°C) **AVERAGE VISIBILITY:** 90 feet (27 m)
AVERAGE DEPTH: 50 feet (15 m) **TYPE OF DIVE:** Wreck and shark

Over the sunken bow of the *Ray of Hope,* a 200-foot-long (61 m) former freighter, the sleek and distinctly sharky silhouette of a Caribbean reef shark glides overhead, momentarily blocking the sun streaming through the sea. With their triangular dorsal fins and arrowhead-shaped caudal fins, Caribbean reef sharks are beautiful to behold—and they like the *Ray of Hope.*

Accessible by boat, the *Ray of Hope*—purposefully sunk in 2003 to be an artificial reef—sits upright on a bed of sugary white sand. It's an ideal wreck for beginners, teeming with angelfish, blue tang, groupers, moray eels, coral, and the reef sharks.

The sharks frequent this wreck because of its proximity to the Shark Arena, which can be dived on the same day as the *Ray of Hope.* Divers descend to a circle of rocks—the Shark Arena—tucking one in between their knees to help them stay grounded on the sandy bottom. Once everyone is in place, the expert shark feeder descends with a bait box. The sharks begin to circle, watching. They know the drill and wait for the cues.

The shark feeder beckons in the swirl of sharks. A controlled spiral of fish eager to get a feed block out the diver in a tornado of fins. Many divers are new to this world. What they leave with is awareness that sharks aren't the mindless hunters they're often portrayed as.

Off New Providence Island you can dive *Ray of Hope* (and the Shark Arena) year-round. Dive only with reliable and responsible operators—ones who will take care of you and respect the sharks.

What You'll See: Caribbean Reef Sharks • Tawny Nurse Sharks • Nassau Groupers • Stingrays • Green Turtles • Green Moray Eels • Angelfish • Blue Tang

Caribbean reef sharks circle
a diver exploring the wreckage.

MICRONESIA

GOOFNUW CHANNEL

Dive Micronesia's manta "car wash."

AVERAGE WATER TEMP: 83°F (28.3°C) **AVERAGE VISIBILITY:** 30 feet (9 m)
AVERAGE DEPTH: 50 feet (15 m) **TYPE OF DIVE:** Open water

Every December through April, Yap's resident population of manta rays (more than 100 individuals) spruce themselves up for mating season, visiting cleaning stations like the one in the Goofnuw Channel, on the northeastern side of Yap.

Divers descend to the shallow channel bottom and stay stationary, watching the spectacle overhead. Mantas approach, usually in twos and threes (sometimes in processions of up to 12), aiming for elevated coral formations where small, specialized "cleaner" reef fish wait to pluck parasites from visitors.

As they circle, the manta rays swoop within inches of divers' heads, performing vast acrobatics with wingspans stretching eight to 13 feet (2.4 to 4 m). This spruce-up can last the length of a 45-minute dive, and those lucky enough to see it never get bored.

If divers can tear their eyes away from the manta rays, they might see whitetip reef sharks resting on the bottom, bumphead parrotfish, snapper, and trevallies. But most have eyes only for the graceful giants, returning to this site again and again.

It does take some time to reach Yap, one of Micronesia's 2,000-plus islands, so it's worth taking a look around while you're here. Yap has a healthy, extensive barrier reef that offers a wide variety of other dive spots, from drop-offs to passes, walls to caverns. Twilight dives are particularly popular. ❍

What You'll See: Manta Rays • Bumphead Parrotfish • Whitetip Reef Sharks • Hard Coral • Trevallies • Snapper • Jacks • Barracuda • Anemones • Clownfish • Ghost Pipefish • Mandarinfish • Nudibranchs • Dragon Wrasses • Fire Gobies

A crab tries to camouflage atop a sea cucumber.

VANUATU

MILLION DOLLAR POINT

A WWII garbage dump turned treasure trove (with a saucy, spiteful history)

AVERAGE WATER TEMP: 83°F (28.3°C) **AVERAGE VISIBILITY:** 65 feet (19.8 m)
AVERAGE DEPTH: Less than 130 feet (39.6 m) **TYPE OF DIVE:** Wreck

h, the absurdity of war. Vanuatu's Million Dollar Point is a wreck dive, but not in the way you might think: It's a million-dollar dump, a collection of hundreds of tons of WWII equipment—bulldozers, jeeps, trucks, tanks, guns, cranes, and even a salvage vessel.

An easy shore dive from Espiritu Santo Island takes divers past a trail of lightly coral-encrusted wreckage lying on a sandy bottom. Studebakers are jumbled together with Willys jeeps, forklifts, graders, and steam rollers, a staggering collection of the machines of war, everything needed to aid the 100,000 Allied troops and support staff once stationed on the island.

The reason for this underwater junkyard? The story goes that, after the war, the Americans knew they wouldn't be able to cart all their equipment back to the United States. They offered to sell it to the joint French-English government for a bargain price.

In what has to be one of the most expensive games of chicken in history, the French-English government turned down the offer, betting they would get the equipment for free when the Americans left the island. Rather than letting that happen, the Americans dumped the entire lot into the sea.

If you can pull yourself away from the wrecks, slow down and admire the macro: Nudibranchs and other small creatures abound in the wreckage and rocky outcrops, more than you'll spot on the nearby S.S. *President Coolidge* (page 260). ❍

What You'll See: WWII Equipment • Rays • Eels • Octopuses • Barracuda • Coral • Hogfish • Nudibranchs

Coral thrives on one of the many WWII wrecks at Million Dollar Point.

HAWAII

KONA

Take a night dive with ocean giants.

AVERAGE WATER TEMP: 75°F (23.9°C) **AVERAGE VISIBILITY:** 22 feet (6.7 m)
AVERAGE DEPTH: 30 feet (9 m) **TYPE OF DIVE:** Night

Graceful as ballerinas, winged giants swoop through the inky black, illuminated by a series of diver-held flashlights that cast a green tinge to the water. Barrel-rolling, backflipping, and banking, manta rays create an underwater dance as they feed on plankton attracted by the lights. It's one of the most mesmerizing spectacles on the planet, a night dive watching a dozen manta rays, each wielding a 20-foot (6 m) wingspan, glide through the water in a way we can never hope to do.

The Kona coast on the Big Island is famous for its resident population of more than 100 individual manta rays, making it one of the most accessible (and healthy) populations. The mantas cruise to the calm Kona waters after sundown to feast on plankton.

For divers new to night diving, many operators offer a two-tank dive, the first taking place in the late afternoon to familiarize divers with the location before dark. (It's also a great opportunity to see the occasional early-bird manta, turning up in advance of the feast.)

Once the sun has set, divers remove their snorkels (so they don't scratch the mantas) and descend to

A pair of manta rays perform a dance for their nighttime visitors.

What You'll See: Manta Rays • Spinner Dolphins • Silky Sharks • Hammerhead Sharks • Tiger Sharks • Humpback Whales • Sperm Whales

sit on the rock-studded bottom with their torches, while snorkelers float flat on the surface, hanging onto a light-studded surfboard fitted with side handles. In the water column between the diver and snorkeler lights, the mantas swoop and feed, sometimes in numbers of just one or two, sometimes up to 20 individuals jockeying for position, an aerial battle of near misses and breathtaking passes.

It's not unusual for divers to be brushed by wingtips, so you'll need to be comfortable adjusting your gear if it gets bumped out of position. Be sure to let the manta dictate the interaction—your underwater role is to sit back, watch, and enjoy, which is one of the reasons this is a good choice for a first night dive.

Manta ray night diving is offered year-round and requires no special certification beyond Open Water. This is an excellent first night dive, although divers should be prepared for gentle surge. If you're unsure, go on the snorkel tour first (take precautions if you're prone to seasickness—it can get rocky on the surface) and watch the dive (as well as the mantas!), then return the following night for the dive.

Located approximately 2,400 miles (3,862 km) from a continental landmass, the Hawaiian archipelago is one of the most remote in the world, with one of the largest proportions of endemic species—about 25 percent. But uniqueness is paired with fragility, and many of Hawaii's native species are under threat. Many have already disappeared altogether.

Being one of the most reliable manta sightings in the world, Kona attracts its fair share of tourists, which can cause problems. Mantas have been injured by boat propellers, and the lights do, of course, influence natural behavior. It is critical that divers do their homework and support quality operators.

On the flip side, the more people who have the opportunity to see and experience these majestic acrobats, the more we're educated, and the better fighting chance this threatened species will have at surviving. If divers and snorkelers are informed about and experience the nighttime underwater dance of the manta rays, they cannot help but be moved—and hopefully moved to care about preserving these magnificent creatures. ❍

Light from a surfboard illuminates a manta ray and schooling fish.

A reticulated brittle star sits atop a cushion sea star.

Yellow tang clean a green turtle's shell.

A diver passes a school of fish on a dive in Kona.

INDONESIA

RAJA AMPAT

Diving fit for a king

AVERAGE WATER TEMP: 82°F (27.8°C) **AVERAGE VISIBILITY:** 30 to 100 feet (9 to 30 m)
AVERAGE DEPTH: 33 to 130 feet (10.1 to 39.6 m) **TYPE OF DIVE:** Open water

Perhaps this is how the world looked at the beginning. Pristine. Prolific. Lush.

Raja Ampat is diving fit for a king—or four kings, which is the translation of the name, referring to the four main islands (Misool, Waigeo, Batanta, and Salawati) in the 1,500-island archipelago off the Indonesian province of West Papua (formerly known as Irian Jaya).

Raja Ampat has been compared to the Amazon rain forest, unrivaled in underwater diversity, one of the planet's richest and most diverse ecosystems. So far, 565 species of hard coral, 1,100 species of reef fish, and 600 species of mollusks have been identified here, with more being cataloged and discovered every day. This is pioneer diving: Dive boats only popped onto the scene in 2002 and—with 15,000 square miles (38,850 sq km) of land and sea to explore—divers haven't even scratched the surface of this wild, underwater Eden.

Accessed by liveaboard or from a handful of land-based resorts, the inviting, aquamarine waters hold a vast and diverse number of dive locations. Boo Rock (also known as Boo Windows or Boo Islands) has to be one of the most familiar and photographed spots, with a large swim-through, schooling fish (unicornfish, fusiliers, and surgeonfish), a coral wall (barrel sponges, *Acropora,* sea whips, gorgonians, and resident nudibranchs), and a few other reef favorites, including octopuses, banded sea kraits, green turtles, titan triggerfish, and batfish.

Cape Kri has a swarm of fish. More than 374 different species were counted on a single dive here. This lazy drift dive sweeps divers past coral gardens exploding with color, with the occasional wobbegong shark relaxing beneath table corals. Farther along, schools of glassfish, sweetlips, blacktip and whitetip reef sharks, dogtooth tuna, Napoleon wrasses, and giant trevallies surpass count.

A plethora of marine life lives among the reefs and leather coral.

A treasure trove for fish spotting, including reef sharks, Napoleon wrasses, and jacks

What You'll See: Sea Fans • Pygmy Seahorses • Jellyfish • Archerfish • Bumphead Parrotfish • Manta Rays • Napoleon Wrasses • Nudibranchs • Moray Eels Jacks • Batfish • Surgeonfish • Fusiliers • Giant Trevallies • Mantis Shrimp • Cuttlefish • Octopuses • Wobbegong Sharks • Blacktip Reef Sharks • Whitetip Reef Sharks • Glassfish • Spanish Mackerel • Dogtooth Tuna

Even the most jaded divers will be impressed by Fabiacet, a collection of pinnacles surrounded by deep, clear water, with high visibility and a prolific collection of fish: hammerhead sharks, fusiliers, green turtles, angelfish, surgeonfish, leather corals, sea fans in resplendent purples, banded angelfish, and triggerfish.

Many scientists attribute Raja Ampat's healthy population to its geographical location, smack in the middle of the meeting place for currents from the Philippines, Australia, and Indonesia. A lack of human impact certainly has something to do with it, too. With few residents and subsistence rather than commercial fishermen, our imprint on this kingly spot is a looming threat, but manageable—if action is taken to preserve Raja Ampat, a proposed UNESCO World Heritage site. ❍

Travel Tip:

Raja Ampat is diveable year-round, with the best months for liveaboards being October through the end of April. July through mid-September can be rougher. Visibility in Raja is surprisingly variable, but it's usually at its best in the mornings. Most of the diving is easy and suitable for beginners, but current and Raja Ampat's remote location means you need to have your wits about you and dive within your limits.

U.S. VIRGIN ISLANDS

SALT RIVER CANYON

Dive an ancient river bed—and legendary wall dive.

AVERAGE WATER TEMP: 81°F (27°C) **AVERAGE VISIBILITY:** 80 feet (24 m)
AVERAGE DEPTH: 20 feet (6 m) to recreational limits **TYPE OF DIVE:** Wall

The Salt River Canyon is hidden in plain sight. This ancient riverbed is one of the oldest geological areas in the Caribbean, a "V" with two vertical walls facing each other across a quarter mile (0.4 km) of deep blue, only a five- to 30-minute boat ride from shore depending on which dive shop you go with.

Each wall has its own distinct personality. The East Wall has a mooring in 40 feet (12 m) of water hanging over a 1,000-foot (305 m) drop-off. The sloping wall makes for an interesting mix of tropical and pelagic species, such as angelfish rubbing fins with hammerheads.

The West Wall begins with a mooring in 20 feet (6 m) of water, before dropping first 200 feet (61 m), then 500 (152), then 4,000 (1,219). There are plenty of large fish to spot here (groupers, jacks, barracuda), but the showstoppers are the geological formations believed to have been carved by a waterfall eons ago, creating fascinating swim-throughs and overhangs for divers to explore.

The Salt River Canyon lies off the Puerto Rico Trench, the deepest part of the Atlantic Ocean, with plunging depths exceeding 27,500 feet (8,382 m). The canyon is legendary for the quality of wall diving, as well as its historical significance. This is the only place in the United States where Christopher Columbus's men are known to have landed on Columbus's second trip to the Americas in 1493. (It didn't go well for the locals.)

Located in the Virgin Islands, the Salt River Canyon is favored with consistent weather, which makes for great warm weather diving. Mind June through October, when hurricanes can be frequent and diving rougher. ○

What You'll See: Sponges • Sea Fans • Horse-Eye Jacks • Black Coral • Hammerheads • Blacktip Reef Sharks • Moray Eels • Spotted Eagle Rays • Angelfish • Parrotfish • Groupers • Barracuda • Old Ship Anchors

The best wall dives can be found within the waters of the U.S. Virgin Islands.

PHILIPPINES

CEBU AND MALAPASCUA

A little bit of everything from schooling sardines to coral gardens to thresher sharks

AVERAGE WATER TEMP: 84°F (28.9°C) **AVERAGE VISIBILITY:** 58 feet (17.7 m)
AVERAGE DEPTH: 70 feet (21 m) **TYPE OF DIVE:** Open water

Cebu is the main island in the Cebu Province of the Philippines, a collection of 167 surrounding islets and islands in the Coral Triangle. Together with Malapascua Island, off Cebu's northern tip in the Visayan Sea, these sites offer dives for every underwater explorer, an explosion of marine diversity.

In Cebu, Moalboal (a spur on the west side of the island) gets top billing. This protected marine park is flush with schools of sardines and barracuda, as well as tunnels and caverns. On the east side, Mactan (a small island close offshore from Cebu City, connected to the mainland by road) is a shore-diving favorite for beginner divers, with reefs teeming with nudibranchs, frogfish, groupers, and turtles leading to walls dropping off in as little as 16 feet (4.9 m) of water.

In Malapascua, Monad Shoal is one of the only places in the world where divers can reliably swim with thresher sharks. They visit Monad Shoal in the early mornings to get a cleaning. Divers set off between 5 a.m. and 7 a.m., either via a stationary line or swim dive, descending to around 70 feet (21 m) overlooking a 700-foot (213 m) drop-off.

These striking creatures—nearly 20 feet (6 m) long, half of which is their namesake scythe-like tail—are mesmerizingly graceful, with a slightly surprised expression, perhaps because they prefer deep water and rarely venture into the shallows. ❍

What You'll See: Thresher Sharks • Cuttlefish • Nudibranchs • Moray Eels • Batfish • Banded Sea Snakes • Frogfish • Scorpionfish • Whitetip Reef Sharks • Mackerel • Pygmy Seahorses • Oriental Sweetlips • Groupers • Pipefish • Mandarinfish

Whale sharks cruise by kayakers and swimmers.

AUSTRALIA

NAVY PIER

One of the world's best jetty dives and an underwater photographer's dream

AVERAGE WATER TEMP: 80°F (26.7°C) **AVERAGE VISIBILITY:** 16 to 65 feet (4.9 to 19.8 m)
AVERAGE DEPTH: 50 feet (15 m) **TYPE OF DIVE:** Jetty

Western Australia's Navy Pier teems with a dazzling concentration of marine life, from inquisitive sea snakes and colorful nudibranchs to toothy sand tiger sharks and giant Queensland groupers.

This 360-foot-long (109.7 m) T-shaped pier is an active Australian Navy facility, and its restricted access works in divers' favor: No fishing and a lack of heavy traffic results in a fishy oasis, with prey attracting predators, an accumulation of species rarely seen concentrated in such a small space.

Accessed by the shore, this shallow dive is easy to navigate (follow the jetty!), even in the variable visibility this area is known for. (See nearby Ningaloo Reef, page 330.) Under the jetty's protection, the ocean goes about its daily life, a dream opportunity for underwater photographers. Octopuses, frogfish, moray eels, and flatworms love the pier's hidey-holes, while wobbegong sharks, snapper, and fusiliers weave in and out of the shadows cast by the structure. With more than 200 fish species calling the Navy Pier home, there is something to delight every diver.

Access to Navy Pier is (understandably) restricted. All diving must be guided by a local licensed operator and all divers must present ID (driver's license or passport) and pay an entrance fee in addition to dive costs. The strict conditions might seem annoying, but they are the reason the Pier is consistently listed as one of the world's best jetty dives, if not one of the world's best dive sites. Plan ahead, roll with the restrictions, and prepare to be amazed. ❍

What You'll See: Barracuda • Trevallies • Sweetlips • Whitetip Reef Sharks • Sand Tiger Sharks • Wobbegong Sharks • Clownfish • Moorish Idols • Blue Angelfish • Whale Sharks (on occasion by season)

An infant blacktip grouper seeks shelter.

HONDURAS

MARY'S PLACE, ROATAN

A palace of dramatic, coral-coated fissures

AVERAGE WATER TEMP: 83°F (28.3°C) **AVERAGE VISIBILITY:** 90 feet (27.4 m)
AVERAGE DEPTH: 90 feet (27.4 m) **TYPE OF DIVE:** Open water

If you're traveling to Honduras and can only choose one dive site, that site should be Mary's Place, a collection of narrow canyons in Roatan reportedly named after the wife of the diver who discovered it. (Legend has it he originally named it "Mary's Crack," but the saucy name was mellowed slightly when it became more popular.)

Many divers feel the site should have been deemed "Mary's Palace," instead of Mary's Place, due to its beauty and impressive structure, exploding with color and coated with coral, and the life it attracts, from seahorses to moray eels.

Volcanic and seismic activity carved up this underwater playground, resulting in three fractures forming narrow canyons 100 feet (30 m) deep. Mary's Place is famous for its coral: Overhangs hum with marine life (spotted drums, creole wrasses, crabs) while feather black coral, sea fans, and wire coral stretch out from the walls, narrowing the passageways dramatically and reaching out to divers as they swim past.

Roatan offers new divers a chance to explore rich canyons and reefs.

What You'll See: Black Coral • Sea Fans • Wire Coral • Barrel Sponges • Creole Wrasses • Crabs • Lobsters • Drumfish • Barracuda • Butterflyfish • Angelfish • Parrotfish • Snapper • Moray Eels

This is where the dive can get tricky—the depths are better for advanced divers with solid buoyancy control. Although beginner divers should avoid descending too far into the crevices to avoid accidentally damaging the coral, there is still plenty to see over the tops of the crevices, around the 40- to 60-foot (12 to 18 m) mark: Large coral heads loom, surrounded by parrotfish, butterflyfish, and angelfish. And with excellent visibility year-round, you won't miss a thing.

For new divers, the first canyon or crevice dive can feel like flying. Walls reach up and around in a familiar way, but you're moving differently. Instead of walking along as you normally would on land, you're gliding, with the ability to move in every direction, and the privilege of seeing in every direction. It's an exhilarating reminder of what diving is all about.

Swim through one canyon, the walls almost within touching distance, before turning into another, and then another—depending on how your air consumption is going, of course. The light takes on a dreamlike quality, illuminating the flash of silversides, or the careful movement of a lobster. This intricate system of fissures and crevices is so full of life it mesmerizes divers, pulling their attention away from the turtle hanging just off the wall, or the squadron of eagle rays gliding overhead. (It also has a habit of distracting divers from checking their depth gauge, so keep your wits about you.)

Mary's Place is located in the south of the island of Roatan, the largest of the Bay Islands off Honduras's east coast and near the towns of Mount Pleasant and French Harbour. The best time to visit is April and May, when the weather is warmest and driest. Most of the rain falls between October and January—but rainy season can begin as early as July, so be prepared.

Due to its popularity, the Roatan does get heavy diver traffic, so expect crowds. Mary's Place has been closed to divers in the past due to fatigue and reef damage from overtrafficking. The site is constantly and closely monitored and managed to make sure the area is protected and preserved. Due to boat restrictions, visitors may have to make drift dives. But the restrictions are for a good cause: The occasional closure gives the ecosystem a breather. When you're diving here, go with the flow—it's for the site's own good. ❍

A channel-clinging crab climbs its home reef.

A brittle sea star climbs over a vase sponge.

Social feather dusters grow in coral crevices.

A grouper swims by for a close-up.

BRAZIL

FERNANDO DE NORONHA

Breeding and feeding grounds for turtles, sharks, and tuna

AVERAGE WATER TEMP: 81°F (27°C) **AVERAGE VISIBILITY:** 160 feet (48.8 m)
AVERAGE DEPTH: 10 feet (3 m) to recreational limits **TYPE OF DIVE:** Shore, open water, wreck, and wall

When you hear of a place rumored to have the best beaches in a country world famous for its beaches, you sit up and take notice.

Fernando de Noronha is an archipelago of 21 islands and islets in the Atlantic Ocean, roughly 220 miles (354 km) from the Brazilian mainland. These sunken volcanoes were designated a maritime national park in 1988, and in 2001 UNESCO made it a World Heritage site, due to its environmental importance.

These healthy waters are breeding and feeding grounds for sharks (lemon, nurse, hammerhead, and Caribbean reef), turtles (hawksbill and green), and tuna, an important haven for marine life in the vast open ocean—and a good spot for divers to know about.

With 24 identified locations, there is something for every diver, from Morro de Fora (a quiet dive ideal for newbies), Caieiras (three rock formations that the turtles like to call home), Caverna da Sapata (a rocky wall covered in sponges and teeming with fish), and Buraco das Cabras, reportedly one of the most colorful spots in a sea of color.

There are even wrecks, like the shallow (23 feet/7 m) *Eleane Stathatos,* or the deeper (197 feet/60 m) and very well-preserved *Corveta Ipiranga.* In these waters, spinner dolphins, pantropical spotted dolphins, humpback whales, and short-finned pilot whales delight divers from both above and beneath the water. ❍

What You'll See: Sharks (Caribbean Reef, Lemon, Nurse, and Hammerhead) • Spinner and Pantropical Spotted Dolphins • Humpback Whales • Short-Finned Pilot Whales • Hawksbill and Green Turtles • Tuna • Barracuda • Squid • Octopuses • Lobster

The Dois Irmãos (Twin Brothers)
peaks on Do Padre beach

TONGA

HUNGA MAGIC

Share tropical water with humpback whales.

AVERAGE WATER TEMP: 81°F (27°C) **AVERAGE VISIBILITY:** 131 feet (39.9 m)
AVERAGE DEPTH: Shallow to recreational limits **TYPE OF DIVE:** Snorkel or free-dive

It's an annual ritual. In mid-July, the juvenile humpbacks show up, circling boats, curious and full of energy. In August, the heat runs begin, with groups of male humpback whales weighing in at 40 tons (36.3 metric tons), stretching 40 to 50 feet (12 to 15 m), chasing females through the warm waters of Vava'u.

Mid-August to mid-September, the males start to sing and calves begin to appear, leading to October, which is the special time for mothers and their calves, luxuriating in Tonga's rich marine environment, resting up for the massive migrations humpback whales are known for. (Some have been clocked traveling 15,000 miles/24,140 km.)

The humpbacks are traveling between their summer feeding grounds in colder water near the poles (feasting on krill, small fish, and plankton) to warmer breeding waters closer to the Equator.

These annual trips between feeding and breeding grounds bring anywhere from 500 to 1,500 whales to Tonga each season, one of the only places in the world where visitors can swim and snorkel with these spectacular—and gigantic—creatures, under the careful guidance of a licensed operator.

"As soon as one of these behemoths comes into view, it's as if time stops. A mix of emotions swirls collectively among us human visitors that can be summed up by two words: sheer awe. The encounter is punctuated by the whales' every song that vibrates through your core."

—SHANNON SWITZER SWANSON, MARINE SOCIAL ECOLOGIST

A humpback whale
and her calf swim by.

A free diver gets a close encounter with a majestic whale.

What You'll See: Humpback Whales • Nudibranchs • Orange Hairy Ghost Pipefish • Giant Clams • Gray Reef Sharks • Banded Sea Snakes • Clownfish • Spotted Eagle Rays • Hawksbill and Green Turtles • Trevallies • Leather Coral • Sea Fans • Black Coral

If you can tear your eyes away from the gentle giants, Tonga also has healthy reefs, tunnels, arches, and walls waiting to be explored.

One spot, the *Clan MacWilliam* wreck, is a steel freighter sitting in 100 feet (30.5 m) of water, a resting place for hard and soft corals, and an ideal hideout for tuna, snapper, batfish, and crustaceans.

Along with humpbacks, Tonga hosts giants of the coral variety. The Coral Garden Reef features coral stretching 12 feet (3.7 m) across, as well as numerous anemones and clownfish, while the popular Kitu Cave is a multichambered cave guarded by six-foot-wide (1.8 m) sea fans. (Go with a guide, who will know which chambers have air pockets and tunnels that are safe to explore.)

From nudibranchs to the rare orange hairy ghost pipefish, to gray reef sharks and banded sea snakes, Tonga is home to a wide array of marine life, both large and small. ❍

Travel Tip:

Composed of more than 170 coral atolls, Vava'u is the most northern of Tonga's three main groups of atolls and islands, 1,200 miles (1,931.2 km) northeast of New Zealand. Tonga has a proud seafaring history, and the friendliness the locals extend to visitors is legendary. The humpback whale season is late July through mid-October.

FRANCE

REUNION ISLAND

Cetaceans to macro life in a flourishing marine reserve

AVERAGE WATER TEMP: 79°F (26°C) **AVERAGE VISIBILITY:** 100 feet (30 m)
AVERAGE DEPTH: Shallow to recreational limits **TYPE OF DIVE:** Reef, cavern, wall, wreck, and open water

Reunion is a lush volcanic island that seems to rise from the Indian Ocean. This French region—a four-hour flight from Johannesburg—is renowned for three things.

The first is vanilla, some of the best in the world. Second is hiking. The number of trails winding around, through, and over this small island is staggering. Much of the hiking is in the shadow of Reunion's two volcanoes—Piton des Neiges and Piton de la Fournaise, the latter still active.

Lastly, Reunion is well known for its diving and wealth of marine life, enhanced by the natural marine reserve created in 2007. Extending along 25 miles (40 km) of coastline, 13 miles (20.9 km) of which is coral reef, the reserve shelters more than 3,500 marine species, including 366 species of soft and hard corals.

The western side of Reunion is settled, with a warm, transparent lagoon, coral gardens, caves, walls, and shipwrecks, most of which are within a 20-minute boat ride and in ideal conditions for beginner divers. In about 50 feet (15 m) of water are the Caves of Maharani, home to large kingfish and lionfish. Cap la Houssaye is a favorite for macrophotographers. Mantis shrimp, ghost pipefish, and nudibranchs make regular appearances here.

For beginners, Le Petit Tombant—at less than 66 feet (20 m) deep—offers pufferfish, barracuda, turbot fish, and stingrays in different environments, from black sand to coral reefs.

And then there are the outer reef drop-offs, home to passing pelagics like bull and tiger sharks, and more than 22 species of cetaceans, including blower, pantropical, and Risso's dolphins, as well as humpback, fin, and sperm whales. ❍

What You'll See: Blower, Pantropical, and Risso's Dolphins • Humpback, Sperm, and Fin Whales • Bull and Tiger Sharks • Sea Urchins • Sea Stars • Parrotfish • Emperor Angelfish • Black Coral • Moray Eels • Surgeonfish • Snapper • Batfish • Groupers • Lobsters

The *Hai Siang* shipwreck lies almost perfectly intact.

FRENCH POLYNESIA

TUAMOTU ISLANDS

Current dive with the shark gods.

AVERAGE WATER TEMP: 80°F (26.7°C) **AVERAGE VISIBILITY:** 150 feet (45.7 m)
AVERAGE DEPTH: Shallow to recreational limits **TYPE OF DIVE:** Lagoon and current

Fair warning—you won't want to leave. The islands that inspired artist Paul Gauguin are as vivid, lush, emerald, and turquoise as your South Pacific dreams. French Polynesia does not disappoint.

South of Hawaii, east of New Zealand, and west of South America, the Tuamotu Islands are one of five island groups in French Polynesia that stud 950,000 square miles (2.5 million sq km) of crystalline sea like gems. Arcing over 930 miles (1,497 km), the 118 atolls that make up the Tuamotu Islands are strung like the pearls that make up most of the Tuamotu Island's economy.

This is a place where sharks are plentiful and revered as gods in Polynesian mythology. Oceanic whitetip, blacktip reef, silky, gray reef, whitetip reef, hammerhead, and tiger sharks are regular diving companions, casting shapely silhouettes as they glide overhead under bright sunlight.

With their open ocean location, the Tuamotus have current, so they're ideal for divers with some experience who are looking to extend their skills. Dive sites are usually pass dives or channel dives, but there are also lagoons to wind down in, or roaring drifts to kick up the adrenaline, all set against colorful coral reefs, and white and pink sand beaches so bright they hurt the eyes.

Fakarava is a large atoll, a sliver of land wrapped around a 15-mile-wide (24 km) lagoon that is also a UNESCO Biosphere Reserve. Dive sites include Tumakohua Pass, a drop into the deep blue, the current drifting divers among schools of Moorish idols, triggerfish, Napoleon wrasses, angelfish, and barracuda. An underwater valley known as Shark's Hole is favored by hammerhead and lemon sharks.

Manihi the most northern atoll in the Tuamotus, is populated with fewer than 1,000 people. Its inner lagoon is used for cultivating pearls (hence its alias: Island of Pearls),

A whitetip reef shark emerges through a school of scarlet soldierfish.

Schooling Pacific double-saddle butterflyfish swarm the warm waters.

What You'll See: Hammerheads • Tiger Sharks • Blacktip Reef Sharks • Lemon Sharks • Oceanic Whitetip Sharks • Gray and Whitetip Reef Sharks • Silvertip Sharks • Manta Rays • Bottlenose Dolphins • Spinner Dolphins • Barracuda • Eagle Rays • Wrasses • Marbled Groupers • Unicornfish

while the outer lagoon is beloved by divers. West Point has far-reaching visibility, all the better to view the pink coral, while the Break is another sharky spot, with gray as well as whitetip and blacktip reef sharks in droves.

Rangiroa ("immense sky") is the largest of the Tuamotu Atolls, and one of the largest atolls in the world. With its massive lagoon, it's also a natural aquarium. The Blue Lagoon—16 feet deep (4.9 m)—is a favorite for reef fish, while the northern passes—Tiputa and Avatoru—are drift dives famous for attracting dolphins.

A perfect mix of lazy lagoons and heart-pumping current, the Tuamotus should be on every diver's list of dream destinations. Plan to spend plenty of time here—and then add on some more. It will be difficult to tear yourself away. Before traveling here, make sure you have some experience in current (up to four knots) and that you've nailed buoyancy control. If you're ready, the shark gods are waiting. ❍

Travel Tip:

Some of the Tuamotu Islands have dive resorts; others are frequented by liveaboards. Expect remote location culture—no credit cards, limited shops, traditional menus. Conditions are best April through November, when plankton attract pelagics.

INDONESIA

MANTA POINT (KOMODO)

Practice drift diving—with manta rays!

AVERAGE WATER TEMP: 80°F (26.7°C) **AVERAGE VISIBILITY:** 33 feet (10 m)
AVERAGE DEPTH: 56 feet (17 m) **TYPE OF DIVE:** Drift

Manta Point, also known as Karang Makassar, lives up to its name. Located southeast of Komodo Island in a nutrient-rich channel between the Indian and Pacific Oceans, this is a manta ray hot spot, an ideal environment the gentle giants use for cleaning, feeding, courtship, and mating. (Pregnant manta rays are often seen here.)

The steady current is what most divers notice first. Although it's strong, this dive is not a difficult one, making it a good first drift dive for the less experienced. Most dive masters use reef hooks to keep divers together, drawing them to the sandy bottom to wait.

As divers settle in, the mantas swoop overhead, blocking the sun. Their individualized spotted bellies, used for identification, make them as recognizable as old friends to experienced dive masters who know their personalities well. Traveling in groups of two or three, and sometimes found in aggregations of up to 80 (the largest congregations are seen between December and February), the mantas are inquisitive, winging closer and closer, often circling for up to 30 minutes if they feel comfortable.

The Indonesian government is making strides here. In 2014 they outlawed manta ray fishing, protecting a species often hunted for their gills, which are used in medicine. It's another positive example of tourism influencing policy, putting a value on these extraordinary creatures, which moves the government to protect them. ❍

What You'll See: Manta Rays • Whitetip Reef Sharks • Blacktip Reef Sharks • Sweetlips • Cuttlefish • Trevallies • Sponges • Eagle Rays

Giant manta rays swim above a free diver.

PART TWO

INTERMEDIATE DIVES

A diver interrupts schooling fish off Costa Rica's Catalina Islands

AUSTRALIA

NORFOLK ISLAND

Isolated, unique, and mostly unexplored

AVERAGE WATER TEMP: 70°F (21°C) **AVERAGE VISIBILITY:** 66 feet (20 m)
AVERAGE DEPTH: 65 feet (19.8 m) **TYPE OF DIVE:** Open water and reef

Norfolk Island is one of the most geographically isolated communities in Australia, which makes it one of the most geographically isolated communities on the planet. Located northwest of New Zealand and east of Brisbane, smack in the middle of the Tasman Sea, Norfolk Island is as unique, odd, and untouched as its volcanic neighbor, Lord Howe Island (page 236), which lies 560 miles (901 km) to the southwest.

Populated with rolling plains, rocky cliffsides, fingernail parings of golden sand beaches, and its namesake pine forests, Norfolk Island has a wild beauty surrounded by a pristine ocean, thanks to a nonexistent commercial fishing industry and lack of freshwater runoff.

The island has attracted—and repelled—a long series of would-be settlers. The Polynesians first arrived around A.D. 1200, then abandoned the island 300 years later (no one knows why). Next, the Europeans established a penal colony here in 1788, and in turn abandoned it in 1814 due to harsh conditions. In 1825 they tried again, with a newer, more brutal penal colony. It was closed in 1855. (UNESCO declared the colony remains a Cultural World Heritage site in 2010.) A year later, Queen Victoria gave Norfolk to the Pitcairn Island–based descendants of the Bounty mutineers, who settled the island—and stayed.

And now Norfolk Island is attracting another wave of travelers: divers.

The island's underwater volcanic landscape is a playground for scuba enthusiasts: Caves, chimneys, tunnels, chasms, and reefs surround it, offering up more than 30 unique dive spots. From protected coral coves to shore dives to open ocean explorations with the big stuff (bronze whaler sharks, grouper, and even the rare whale shark), Norfolk Island and its sister islets—Phillip and Nepean Islands—are all part of the protected Norfolk Island National Park. And although fishing is popular with the locals, no commercial fishing takes place here, resulting in huge schools of trevally, kingfish, and Norfolk

An Australian yellowtail
kingfish faces a diver head-on.

The rocky beaches of Norfolk Island

What You'll See: Groupers • Bronze Whaler Sharks • Galápagos Sharks • Hammerheads (on occasion) • Kingfish • Rock Cod • Nudibranchs • Flatworms • Trevallies • Snapper • Red Emperor • Hawksbill and Green Turtles • Sea Stars • Norfolk Nanwi

nanwi, also known as the "dream fish." (Apparently eating nanwi can cause nasty hallucinations.)

Duncombe Bay on the north coast, a popular location with hawksbill turtles, has a long, underwater wall studded with caves and swim-through arches. On the east coast, Steels Point has three pinnacles of rock favored by schooling fish, and an ocean floor of sea stars.

The Gun Club at Anson Point is home to a coral-covered wall, large underwater boulders, and swim-throughs filled with rock cod and snapper.

Most of the dive sites experience currents and surge; some are high tide only, while others have underwater formations, which is why divers should visit Norfolk Island with some experience under their weight belts to make the most of the experiences this unique and isolated island has on offer. ❍

Travel Tip:

Flights operate four times a week to Norfolk Island from Sydney on Fridays and Mondays, and from Brisbane on Saturdays and Tuesdays. Once on Norfolk Island, a rental car or bicycle is required to get around (watch out for livestock!). Although English is Norfolk Island's official language, you will hear plenty of the local language, a lilting mix of Tahitian and 18th-century English, which the Bounty descendants brought over.

SOLOMON ISLANDS

KAVACHI CORNER

Dive within earshot of one of the most active submarine volcanoes in the world.

AVERAGE WATER TEMP: 84°F (28.9°C) **AVERAGE VISIBILITY:** 115 feet (35.1 m)
AVERAGE DEPTH: 95 feet (29 m) **TYPE OF DIVE:** Open water

Often you feel the volcano before you hear it. A rattling in your teeth. A rumbling below your breastbone. An ominous sense of lurking power.

Located in the western Solomon Islands, Kavachi is famous for two things: The first is its destructive ability to both create and tear itself to pieces, violently and frequently. Since its first recorded eruption in 1939, Kavachi has formed and eradicated a new island at least eight times. Known locally as Rejo te Kavachi (Kavachi's Oven), it's not a stretch to imagine a restless and willful ocean god fretting and fuming beneath the waves, chafing at mortal confinement.

Kavachi also made headlines in 2015 when a National Geographic Expedition discovered sharks living *inside* the crater. The footage of hammerheads

Travel Tip:

Come equipped with a sense of adventure and speak with local operators about weather conditions before booking. The site is virtually inaccessible during rough seas. November generally provides the best conditions.

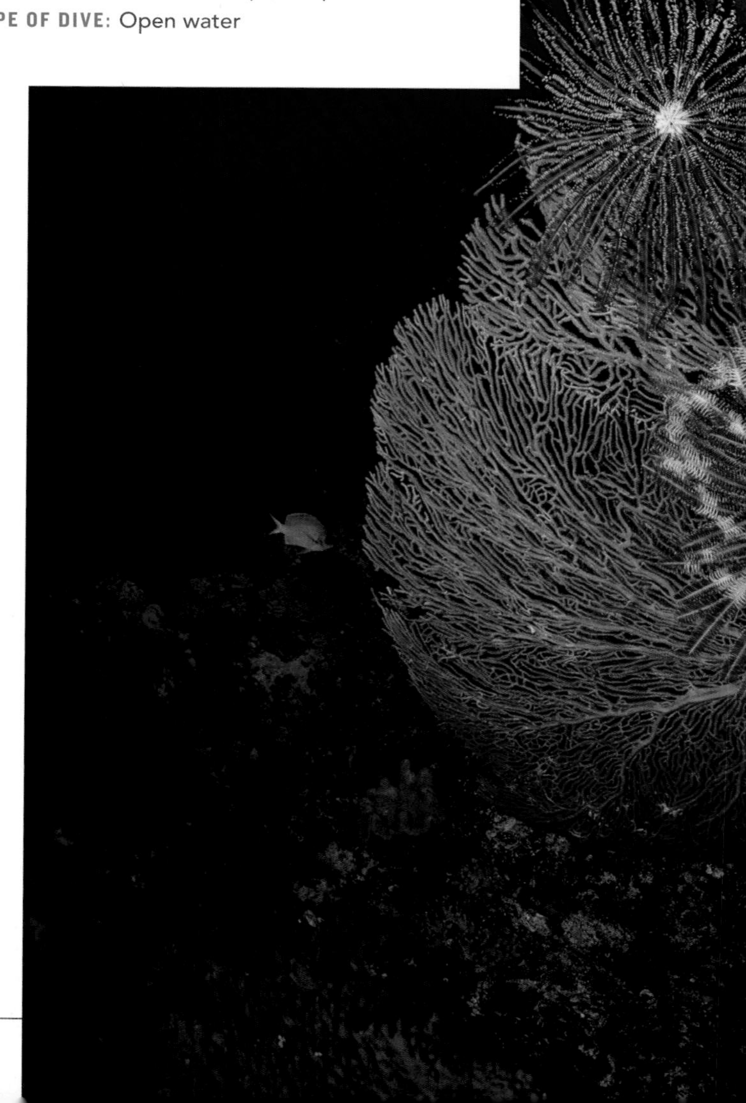

The warm waters offer the ideal habitat for feather stars and sea fans.

What You'll See: Rays • Giant Trevallies • Mackerel • Bumphead Parrotfish • Silvertip Sharks • Bigeye Jacks • Clark's Anemonefish

and silky sharks cruising around the boiling, acidic environment went viral, quickly garnering the nickname "Sharkcano."

The team had chosen Kavachi to study geothermal activity, lowering their instruments at a time when the volcano was strangely quiet. What they found was completely unexpected: jellyfish, snapper, a sixgill stingray, and then the unmistakable shape of a scalloped hammerhead, along with a silky shark, their presence revealing just how little we know about the deep spaces in the ocean. There be gods down there, with sharks as watchdogs.

Divers, obviously, would be unwise to venture too close to Kavachi's Oven, but you can dive close enough to feel the sea god's presence. Kavachi Corner is a spot 18 miles (29 km) south of Kicha Island, northeast of the volcano, an open water dive frequented by spawning giant trevallies and mobula rays.

Approximately 15 miles (24 km) from the (currently) submerged volcano, it's a good spot to shake some fillings loose. The currents can be strong here, and Kavachi makes its presence known from time to time, a rumble that courses throughout your body. (It's reported that the tremor can be felt up to 50 miles/80.5 km away.)

Located in the Marovo Lagoon in the Strait of New Georgia, the visibility in Kavachi Corner is excellent—averaging 115 feet (35.1 m), depending on current and tide—all the better to see schools of fish—bumphead parrotfish by the dozen, mackerel, and spawning giant trevallies, as well as mobula rays.

The Solomon Islands are renowned for a wide variety of dive locations, from WWII wrecks, to mask-ripping current dives (see Devil's Highway, page 366), to underwater volcanoes. Due to its spread-out nature and lack of infrastructure, liveaboards are a good way to explore what the Solomons have to offer, including Kavachi Corner. ○

"Where currents and worlds collide. I was surrounded by hundreds of different fish species at once; I have rarely felt more like just another fish myself."

—ANDREA REID, FISH BIOLOGIST, DIVER, AND CONSERVATIONIST

A diver explores the Mbuco Caves in the Marovo Lagoon.

Lionfish patrol the reefs.

A pair of clownfish swim in and out of their anemone home.

The turquoise and sapphire Marovo Lagoon

CYPRUS

M.S. *ZENOBIA*

A large and unusual wreck with suspended vehicles

AVERAGE WATER TEMP: 72°F (22.2°C) **AVERAGE VISIBILITY:** 65 feet (19.8 m)
AVERAGE DEPTH: 97 feet (29.6 m) **TYPE OF DIVE:** Wreck

One year after her maiden voyage, the M.S. *Zenobia* capsized and sank near Cyprus. Launched in 1979, this Swedish-built ferry now lies on her port side, in depths ranging from 52 to 141 feet (15.8 to 43 m).

The M.S. *Zenobia*'s cavernlike hold offers an unusual penetration dive for those with the experience and training. When the ship sank, it slipped below the surface with more than 100 vehicles chained to its cargo deck. A few of those vehicles still remain, suspended, surreal. (Fortunately, all passengers were safely rescued and there were no casualties.)

The wreck lies near Larnaca, Cyprus, roughly a mile (1.6 km) from shore, requiring a brief boat ride to reach the dive site. "The *Zen*" (as it's affectionately called by locals) is larger than most divers expect: 590 feet (179.8 m) long and 92 feet (28 m) wide, surrounded by trucks that litter the seabed. Most divers explore the spot in two dives.

The first dive follows the ship along the hull to the stern, past the *Zen*'s imposing propellers (keep an eye out for moray eels in and around the propellers), returning along the ship past the still intact lifeboats. The second dive traces the bow along the bridge to the accommodation block, sharing the space with triggerfish, barracuda, and the occasional tuna.

Conditions of the *Zen* are deteriorating, which is why—coupled with the depth and swim-through opportunities for divers with experience—it's classified as a more challenging dive. The ship is still holding together, but sooner is better than later if you'd like to dive the *Zen*.

The diving season runs from March through November, which is the longest in the Mediterranean. Cyprus seems to miss out on the winter storms that can whip and roil the rest of the Med, so take advantage! ○

What You'll See: Wreck • Groupers • Barracuda • Nudibranchs • Tuna • Triggerfish

An overturned truck lies amid WWI wreckage.

Crinoid coral and feather stars shelter fish in the reef.

FIJI

SOMOSOMO STRAIT

The soft coral capital of the world

AVERAGE WATER TEMP: 79°F (26°C) **AVERAGE VISIBILITY:** 90 feet (27 m)
AVERAGE DEPTH: 75 feet (22.9 m) **TYPE OF DIVE:** Open water and reef

The Somosomo Straight is a squeeze of a channel between Fiji's second and third largest islands, Vanua Levu and Taveuni. This nutrient-rich current feeds the 19-mile-long (30.6 km) Rainbow Reef. This is an underwater dream, with everything from whitetip reef sharks, barracuda, and zebra sharks, to delicate gorgonians and feathery lionfish.

The Great White Wall is the premier site, and one that lives up to its legendary reputation. Divers descend to 50 feet (15 m), past corals, sponges, and crinoids, into a swim-through, and then they see it: a vertical wall in coral so white it glows like ice.

Everything depends on timing: Hit the tide and current right and the coral changes color, dark to white, as if blanketed with a sudden snow.

What You'll See: Giant Trevallies • Zebra and Whitetip Reef Sharks • Moray Eels • Blue-Spotted Rays • Manta Rays • Ribbon Eels • Pygmy Seahorses • Nudibranchs Clownfish • Gorgonians • Staghorn Coral • Christmas Tree Worms • Triggerfish

A mantis shrimp *(Squilla empusa)*

It's dramatic, breathtaking, and mesmerizing. (Watch your depth—it's easy to get distracted and drift deeper than you planned here.)

At Rainbow Reef, bommies bloom with soft corals in an array of colors, festooned with brightly hued Christmas tree worms reminiscent of something found in a Dr. Seuss book. Anemonefish and triggerfish also patrol the area.

Looking for something bigger? The Fish Factory is known for schools of pelagic fish—trevallies, Spanish mackerel, and barracuda—while whitetip reef sharks rest on the bottom or lazily cruise past bommies populated by moray eels and dive-bombed by damselfish. Beginner divers can hang out around the 30-foot (9 m) mark with plenty to see, while more advanced divers can drop down the slope to 66 feet (20 m) where there are those pelagic fish in staggering numbers.

Travel Tip:

The best way to experience the Somosomo Strait is to stay at one of the resorts on Taveuni or Vanua Levu for several days. April through October is the primary diving season (and the best visibility), while November to March has the most rain.

The Stairs extends from the southwestern tip of Taveuni, featuring walls that seem to stretch in every direction, punctuated by swim-throughs. Corals cling to the walls, while the distinctive silhouette of a zebra shark pulls your attention away from the parrotfish, fusiliers, and other reef dwellers.

The beauty of the Somosomo Strait is its movement, tidal changes that can also catch divers by surprise, or even a current showing up where there wasn't one before. This is where local knowledge comes into play, so soak in every tip and hint you can find—and dive with an expert guide for safety. ❍

A brittle star feeds on red gorgonian coral.

A small village sits on the shores of the Somosomo Strait.

BOLIVIA AND PERU

LAKE TITICACA

Dive into one of the highest lakes in the world.

AVERAGE WATER TEMP: 54°F (12.2°C) **AVERAGE VISIBILITY:** Variable
AVERAGE DEPTH: Up to recreational limits **TYPE OF DIVE:** Lake

Lake Titicaca is an ancient lake, one of less than 20 that remain on Earth, thought to be three million years old and a UNESCO World Heritage site. South America's largest freshwater lake sits at an altitude of 12,500 feet (3,810 m)—the highest of the world's large lakes—between Peru (west) and Bolivia (east). This is the world's highest dive site within recreational limits—literally breathtaking.

Divers who want to try their hand at this unique and ancient spot need to go through the Bolivian Navy's high-altitude diving center to receive a diving brief, as well as a test in the center's decompression chamber.

The altitude makes this dive more challenging—that and the fact that it's still a relatively unknown dive destination. However, for intrepid divers, this is an ancient place, and you have the privileged opportunity to be a part of its story.

Framed by the snowcapped Andes, this indigo lake has been a gathering spot for humans since time untold. Its cold waters have revealed gold statues, pottery, terraced vertical farms, even an entire underwater building. It is also home to a surprising number of endemic aquatic species such as killifish and catfish—a treasure trove of a different kind.

If you're lucky, you'll spot the holy Kaira frog, which can only be found in these waters. Be on the lookout because a spotting of this amphibian is rare. These larger-than-life frogs are critically endangered due to overhunting, predation, and habitat destruction.

More than 25 rivers empty into Lake Titicaca, so expect slightly brackish water. The best time to visit and dive the area is May through October, when you'll find warmer and sunnier days. The nights still plunge to near-freezing, so pack accordingly. ❍

What You'll See: Artifacts • Holy Kaira Frogs • Aquatic Sponges • 23 Species of Killifish • 2 Species of Catfish

Titicaca is the world's highest diveable lake.

PALAU

BLUE HOLES & BLUE CORNER

Cinematic lighting illuminates congregations of fish.

AVERAGE WATER TEMP: 84°F (28.9°C) **AVERAGE VISIBILITY:** 88 feet (26.8 m)
AVERAGE DEPTH: 59 feet (18 m) **TYPE OF DIVE:** Open water, reef, and cavern

The western Pacific country of Palau is a paradisiacal scene straight off a 1960s postcard: palm trees shading sugary white sand beaches against a palette of blues and greens so bright it hurts the eyes. However, with reefs, walls, caves, channels, and wrecks all bursting at the seams with more than 1,300 species of fish—including Napoleon wrasses that grow to be six-foot (1.8 m), 300-pound (136 kg) mammoths—the paradise of Palau might very well be underwater, rather than above.

The Blue Corner and Blue Holes are two of Palau's most exquisite spots.

Located west of the island of Koror, the Blue Holes sit slightly north of the Blue Corner. The first entrance is in the shallows of the reef, dropping down to the cavern floor 120 feet (36.6 m) below. Take turns descending with your dive buddy, so you can capture each other in one of the most iconic underwater shots you'll ever get: the silhouette of a diver, suspended in blue, perfectly framed by the rock walls of the surrounding cavern, lit from above with the white light of the sun.

Once you've made your descent, whip corals cling to the walls, as schools of fish (barracuda, fusiliers, and jacks) flash by. Divers need to be comfortable with overhead environments here, and those with the proper experience will enjoy exploring nearby caverns and caves.

The nearby Blue Corner is a triangular terrace sticking out of the reef at a depth of 66 feet (20 m), surrounded by walls and hemmed in by congregations of fish—barracuda, whitetip reef sharks, gray sharks, bigeye trevallies, and jacks. Even the most well-traveled and experienced divers are gobsmacked by the sheer quantity of fish carousing in the indigo. Not every dive yields every species that lives here, but every dive will reveal

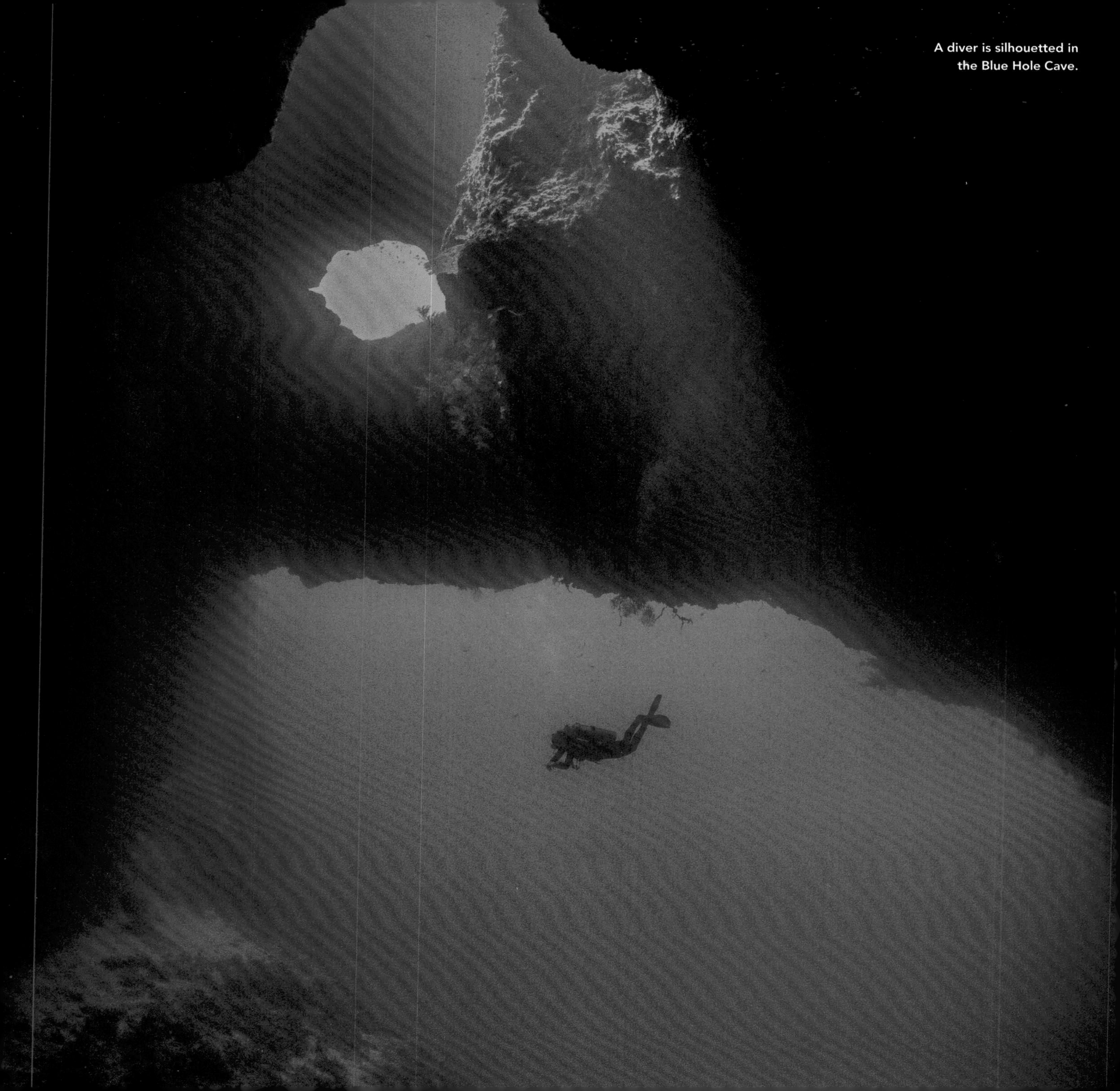

A diver is silhouetted in the Blue Hole Cave.

Millions of jellyfish without stingers take over aptly named Jellyfish Lake.

What You'll See: Napoleon Wrasses • Gray, Zebra, Whitetip, and Blacktip Reef Sharks • Hammerheads • Barracuda • Bigeye and Giant Trevallies • Horse-Eye Jacks • Fusiliers • Tuna • Sailfish • Green and Hawksbill Turtles • Triggerfish • Scorpionfish • Nudibranchs • Sea Fans • Wire Coral • Black Coral

something spectacular, which is why you'll want to return to it again and again.

The trick to this dive is the current, which is what brings in the large numbers of fish. Most dives involve a wall drift past gorgonians and soft corals until you reach a spot where a reef hook will help you pause and enjoy the view. All too soon it will be time to release, cruising to a 33-foot-deep (10 m) plateau favored by Moorish idols, butterflyfish, and red-toothed triggerfish.

Plan your dive carefully and you can do both of these sites together, entering the Blue Holes and using the currents to wind up at the Blue Corner. But you have to keep a careful eye on your depth and air consumption—and be prepared to swim against the current for 150 feet (45.7 m).

Travel Tip:

Book well in advance for a liveaboard on this year-round diving destination—it's hugely popular. November through May offer the best conditions. Make sure to plan a detour to Jellyfish Lake, on the island Eil Malk. Many moons ago, moon and mestiga jellyfish were trapped in the lake. With no predators, they evolved without stingers. Snorkeling surrounded by millions of ethereal, floating formations is an experience not to be missed.

The nutrient-rich waters of the Blue Corner attract hundreds of species of fish—and divers.

A green turtle glides by.

The Blue Holes and Blue Corner have been described as a champagne glass ("too many bubbles"), due to its popularity with divers. But that's a good thing—it highlights the importance of wild places and of protecting them.

Palau paid attention. In 2009, the country announced that ecotourism, not industrial fishing, is their way forward. The same year, they created the world's first shark sanctuary, banning commercial fishing in a 230,000-square-mile (595,700 sq km) protected area. And in 2017, the island nation became the first country to institute a mandatory pledge for tourists—stamped right onto their passports. As President Tommy Remengesau, Jr., noted, environmental stewardship is a Palauan value, and it is important that the nation's guests share in that understanding. A portion of the pledge reads: "I vow to tread lightly, act kindly and explore mindfully. I shall not take what is not given. I shall not harm what does not harm me." Words all divers can—and should—live by. ❍

AUSTRALIA

OSPREY REEF

A submerged atoll with vertical reef and plenty of sharks

AVERAGE WATER TEMP: 79°F (26°C) **AVERAGE VISIBILITY:** 130 feet (39.6 m)
AVERAGE DEPTH: 98 feet (29.9 m) **TYPE OF DIVE:** Open water

Perched atop a seamount in deep, deep water, Osprey Reef—located northeast of Queensland—stretches out over 75 square miles (194 sq km) of cerulean blue sea. It's one of the most northerly reefs in the Coral Sea, 217 miles (349 km) off the east coast of Australia. This submerged atoll is one of the premier Great Barrier Reef destinations, renowned for its healthy reef quality, visibility, and healthy shark populations.

It's liveaboard-only to get here, with a feeling of remoteness that melts away the minute you descend under the waves. Soft corals bloom more than six feet (1.8 m) in height, blocked out by shoals of silver fish. There are more than 15 dive sites around an oval-shaped lagoon 98 feet (29.9 m) deep that drops off within a half mile (0.8 km) to more than 3,200 feet (975.4 m) deep, creating an amphitheater of marine action.

The North Horn spot is Osprey's jewel, affectionately known as shark central. There are several ways to dive this extraordinary spot. When current is present, divers can drift along, or dive deep up to recreational limits. However, it's the shark feed that gives the North Horn its reputation.

Shark feed dives usually begin with an orientation. Divers descend 100 feet (30 m) to a bommie covered in sea fans and whips, where they will watch eagle rays, keeping an eye out for the missed-in-a-flash sailfish and the sharks that usually make an appearance prior to the feed.

When it's time for the actual shark feed dives, divers descend and seat themselves in a 40-foot-deep (12 m) natural amphitheater with a floor of coral debris while guides release fish heads from a sealed drum (operating under strict guidelines). Up to 30 sharks (whitetip reef, silvertip, gray reef, silky, and even the occasional hammerhead) cruise in a pecking order pattern known only to themselves. Large potato cod join in,

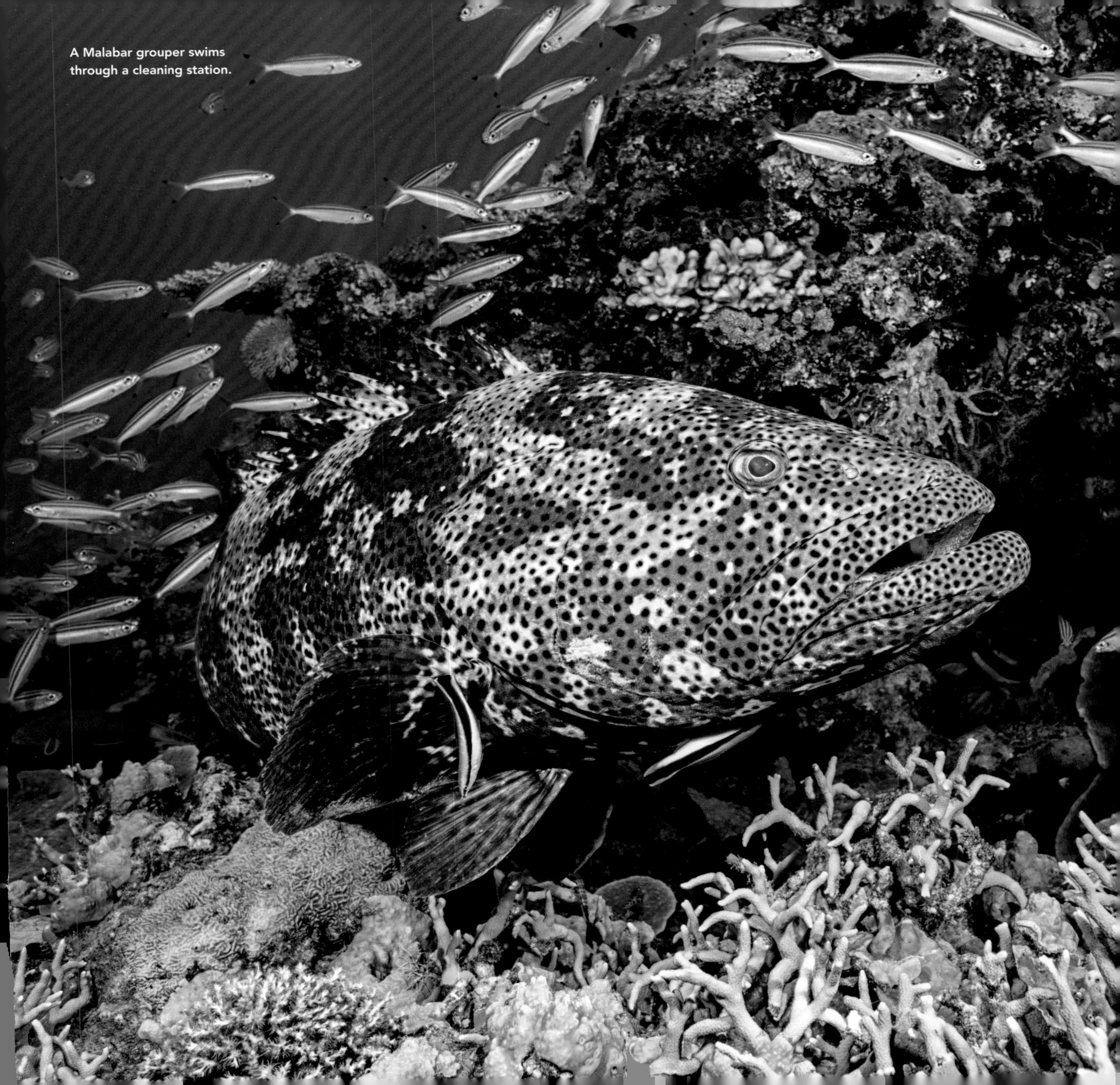

A Malabar grouper swims through a cleaning station.

Osprey Reef, seen from above

What You'll See: Sea Fans • Green and Loggerhead Turtles • Whitetip, Silvertip, Gray Reef, and Silky Sharks • Eagle Rays • Manta Rays • Potato Cod • Fusiliers • Red Bass • Whale Sharks • Beaked and Sperm Whales • Bottlenose Dolphins • Sailfish • Marlin

while smaller fish (including fusiliers and red bass) dart in for scraps.

The False Entrance is another popular site. Its name derives from its mimicry of the actual (and safe) entrance to the lagoon, a few miles to the north. Done as a drift dive in current and as a normal reef dive during slack water, the visibility gives the drop that much more punch, and the array of life (hawkfish, banded pipefish, stonefish, scorpionfish, trevallies, and barracuda) means there is always something to look at.

Osprey showcases everything divers want from the Great Barrier Reef: flourishing marine and coral life (large and small), walls and channels, clear visibility, and warm water. This is the most massive living structure on Earth and, therefore, one of the most vulnerable. By traveling to the Great Barrier Reef, divers confer a value onto the reef, which

Travel Tip:

This is a liveaboard destination only, so book well in advance. Osprey Reef is diveable all year round, although different seasons bring different things: December through February is the rainy season, June through November is a good time to spot humpback whales, while the coral spawning usually takes place in October and November.

A chambered nautilus floats by red whip coral.

A lionfish hunts along the ocean floor.

thereby helps protect it. The Great Barrier Reef has never been under greater threat. The reef is under pressure: In the past three decades it has lost nearly half its coral cover due to human impact and climate change. Diving tourism brings economic resources to this underwater treasure and inspires legislation to protect and heal the reef. There has never been a better time to visit. ❍

"Osprey is a remote isolated oasis known for steep walls, colorful forests of soft coral, pelagic encounters, and a shark dive at North Horn."

—DAVID DOUBILET, UNDERWATER PHOTOGRAPHER

CHILE

EASTER ISLAND

Dive Rapa Nui's famous submerged moai statue and brilliant blue waters.

AVERAGE WATER TEMP: 70°F (21°C) **AVERAGE VISIBILITY:** 120 feet (36.6 m)
AVERAGE DEPTH: 100 feet (30 m) **TYPE OF DIVE:** Open water

Most people wouldn't expect a tiny, isolated island smack in the middle of the Pacific Ocean (the closest landmass, Chile, is 2,300 miles/3,701.5 km away) to have calm seas and surrounding waters of such a brilliant sapphire blue that visibility can reach 200 feet (61 m). But everything about Easter Island is a bit of a surprise.

Rapa Nui, the indigenous name of the place, developed from a Polynesian origin in near isolation, creating a culturally rich landscape best represented by the gigantic, carved stone figures known as *moai,* which are still a powerful attraction for visitors to this 64-square-mile (165.8 sq km) island and UNESCO World Heritage site.

If seeing the moai on land gives you chills, imagine seeing one underwater. In the bay of Hanga Roa, 68 feet (20.7 m) below the brilliant blue surface, lies the inscrutable face of one of these stone behemoths. It's a movie prop replica, not an original moai, purposefully sunk to be an artificial reef and attraction for divers. Still, its effect is instantaneous—and fun, which is what diving should be. Every diver wants to have his or her picture taken with this Easter Island icon.

The submerged moai isn't the only reason Rapa Nui attracts divers from around the world. Its crystal clear waters are famous—there are no rivers on Easter Island, and any rain runoff is filtered through volcanic rock. There isn't a port here, either. The result is spectacular visibility, the better to see green turtles, moray eels, and a structured landscape of hard corals. Positioned on a submerged volcanic ridge, Rapa Nui's ocean landscape is flush with rock formations, swim-throughs, caves, and lava tubes.

Divers are taken to sites around the island in traditional fishing boats and need to be comfortable taking their gear on and off in the open water. Most dives take place on

Long-spined sea urchins cling to coral rocks.

The famed underwater moai replica attracts hundreds of divers a year.

What You'll See: Green Turtles • Kotea Hiva • Angelfish • Easter Island Spiny Lobster • Trumpetfish • Boxfish • Barracuda • Easter Island Eel • Balloonfish

the west coast, due to prevailing weather, or the three *motus* (islets) on the southwest coast if the weather permits.

Marine life, however, is thin. Although nearly one-fifth of the fish species are endemic, Rapa Nui is in desperate need of further protection to safeguard these species, as well as replenish the tuna stocks and shark populations that appear in Easter Island's history. (Ancient petroglyphs of tuna and turtles can be found on the island, and sharks feature in oral legends.) There are community discussions around fishing practices and the need for protection, but the discussions are slow moving, further hampered by the roughly 2,300 miles (3,700 km) between Easter Island and Chilean administration.

Even without the flourishing marine life that we know is possible, this is one of the best—and most unique—dive locations in the world. Add a well-maintained marine reserve and Easter Island would be the envy of the underwater world. ❍

Travel Tip:

Easter Island has plenty of dive centers to choose from; just keep in mind that some are closed on Sundays. Winter offers the best diving conditions and fewer tourists. On the boat ride to your dive site, take a moment to appreciate that the Polynesian society who flourished here did so after arriving in wooden outrigger canoes—a navigational feat if there ever was one.

AUSTRALIA

CHRISTMAS ISLAND

From the red crab migration to limestone caves, this unique island is uncrowded and unusual.

AVERAGE WATER TEMP: 82°F (27.8°C) **AVERAGE VISIBILITY:** 100 feet (30 m)
AVERAGE DEPTH: 60 feet (18 m) **TYPE OF DIVE:** Open water, reef, wreck, and cavern

Places that are difficult to reach usually yield great rewards, and Christmas Island is no different. This Australian territory, located 1,600 miles (2,575 km) northwest of Perth and 310 miles (499 km) south of Jakarta, Indonesia, has the looks and offerings to tempt tourists, but its remote location means that it's still flying under the radar.

The one event that has garnered this island plenty of attention is the annual red crab migration, which happens between October and November at the beginning of the wet season. The red crab is a Christmas Island original—it's not found anywhere else in the world, so this annual migration of nearly 120 million crustaceans is a pretty spectacular site. Their journey is tied to a specific lunar schedule. When the time is right, the crabs leave their solitary rain forest burrows and head to the beach to breed; a blood red tide consumes Christmas Island, flooding roads, paths, and every available space. (It's so plentiful it requires road closings and man-made crab tunnels.) This migration is very specific: If there's a weather delay, the crabs will postpone their festivities until the next lunar month.

When the crab party abates, Christmas Island quiets down and dips off the radar of all but the most curious travelers, who are quickly rewarded for their efforts to get here. More than half of Christmas Island is national park, a lush wilderness ideal for exploration on foot, yielding spectacular lookouts, silvery waterfalls, and secluded beaches. There are even limestone caves to explore.

If possible, there's even more to see underwater—more limestone caves, reefs, walls, large pelagics, and a well-preserved wreck from World War II. This is a spot for more experienced divers: Some locations feature strong currents whereas others are simply more remote.

Located on the tip of an extinct volcano near the Java Trench, Christmas Island is comprised of limestone and uplifted and ancient coral reefs. A narrow reef circles the island

The famous red crabs climb the rocky outposts along the shore.

Pyramid butterflyfish take shelter under a table coral at Christmas Island.

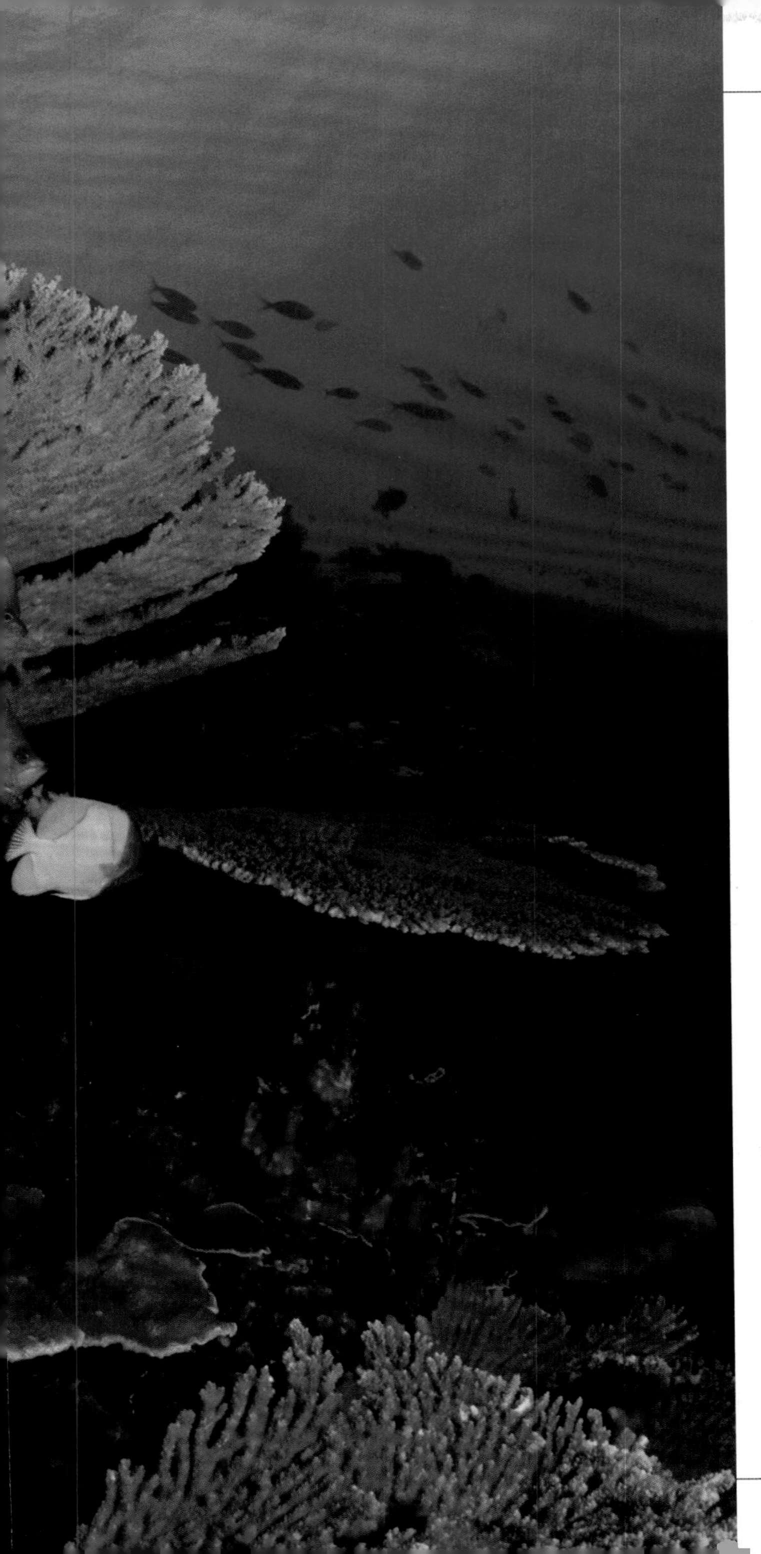

What You'll See: Tiger, Gray Reef, and Silky Sharks • Hammerheads • Pilot and Beaked Whales • Dolphins • Trevallies • Snappers • Tuna • Barracuda • Stingrays • Eagle Rays • Manta Rays • Sea Fans • Angelfish • Butterflyfish

before plunging into the depths, creating an ideal landscape for both reef fish and large pelagics, and divers who have more than 60 unique venues to choose from.

Thundercliffe Cave, for example, is two caves in one. A large open chamber leads past butterflyfish, angelfish, groupers, and barracuda to a second cave filled with stalactites and stalagmites. Divers can even slip off their gear and explore this cave on foot before returning to the water.

The *Eidsvold* wreck, a Norwegian phosphate ship sunk by a Japanese submarine in WWII, lies between 15 and 60 feet (4.6 and 18 m) of water, with a healthy growth of coral and plenty of tropical fish to look at.

And then there is the aptly named Perpendicular Wall, falling vertically, studded with large sea fans and zooming reef fish.

Whatever your pleasure—large, small, or unusual—you'll find something to interest you around Christmas Island. It's worth the migration to experience this place. ❍

Travel Tip:

Christmas Island is diveable year-round, with calm conditions most of the year except during the rainy season, which is November through March. This is also the time that whale sharks and manta rays make an appearance to enjoy the plankton.

MEXICO

GORDO BANKS

Find yourself mesmerized by schooling hammerheads.

AVERAGE WATER TEMP: 73°F (22.8°C) **AVERAGE VISIBILITY:** 60 feet (18 m)
AVERAGE DEPTH: 120 feet (36.6 m) **TYPE OF DIVE:** Open water

On a one-hour boat ride from San Jose del Cabo, at the tip of the Baja Peninsula, the sea can be heavy, and there's an immediate and overwhelming sense of the open ocean, and just how powerful it can be.

Lying nine miles (15 km) offshore is Gordo Banks, a dive site with two submerged seamounts—the tip of one reaches 120 feet (36.6 m) below the waves, the other is 180 feet (55 m) deep. Divers have to be ready to go. Due to the current and surface conditions, there isn't any time for gear adjustments at the surface.

During the descent, the effort immediately begins to pay off. Swirls of large fish—yellowfin tuna, barracuda, groupers, mackerel, amberjacks, and snapper—flash to depths out of sight. This is ray city: eagle, mobula, cow-nose, diamond, butterfly, electric—even manta rays—are frequently spotted at Gordo Banks, drawn by the nutrient-rich currents.

But the reason divers brave the crunching current, depth, and variable visibility in Gordo Banks is the hammerheads. December through May might have cooler surface (and water) temperatures, but the pelagics prefer it, and it's the best time to see hammerheads.

It's never a guarantee, but on the right day, hammerheads school here in mesmerizing numbers, their iconic silhouettes effortlessly swimming above, below, and near, an endless array of charcoal-colored triangles framed against the deep blue of the ocean.

Choose your operator carefully: The switched-on ones will request any diver with fewer than 100 logged dives to do a local two-tank dive in an easier spot prior to visiting Gordo Banks. Although Gordo isn't a hard dive, it is one where divers need to feel confident. ❍

What You'll See: Hammerheads • Silky, Whitetip Reef, Bull, and Tiger Sharks • Mobula, Eagle, Cow-Nose, Diamond, and Butterfly Rays • Whale Sharks • Nudibranchs • Black Coral • Yellowfin Tuna • Barracuda • Groupers • Amberjack • Snapper • Mackerel

A hammerhead shark glides in the waters.

NEW ZEALAND

FIORDLAND

Sharp teethlike mountains framing fjords with 30 feet (9.1 m) of freshwater layered on the sea

AVERAGE WATER TEMP: 59°F (15°C) **AVERAGE VISIBILITY:** 39 feet (11.9 m)
AVERAGE DEPTH: 60 feet (18 m) **TYPE OF DIVE:** Open water

Tucked away in the southwest corner of New Zealand's South Island, Fiordland National Park is a land of extremes, with 15 fjords plunging 1,300 feet (396 m) deep and toothy peaks reaching 5,522 feet (1,683 m) into the sky. This 2,965,265-acre (12,000 sq km) UNESCO World Heritage area is an ethereal, otherworldly landscape like you've never seen before, with gray mountains partially covered with dark green scrub rising sharply from oily-looking, still jade waters, a breathtaking Jurassic landscape amplified.

It rains 25 feet (7.6 m) a year here. Twenty-five feet.

But that rain brings a boon: First, the silvery waterfalls that thread their way down the steep faces of the mountains, diamond veins that make every drop of precipitation a precious gift.

Travel Tip:

There is no dry season to worry about. Expect rain all year round, but that doesn't affect the diving. If you're crowd-averse, avoid New Zealand summer (December through February) when the road to Milford Sound is clogged with campervans.

A diver swims through a black coral forest.

What You'll See: Black, Red, and Stony Coral • Sea Pens • Glass Sponges • Snake Stars • Rock Lobster • New Zealand Fur Seals • Octopuses • Fiordland Crested Penguins • Bottlenose Dolphins • Tube Anemones • Leatherjackets • Butterfly Perch • *Jason mirabilis* Nudibranchs • Brachiopods (aka lamp shells)

Second, the unique dive site. That rain strips tannins from the soil, layering it 30 feet (9 m) thick onto the saltwater in the fjords. It chokes off the light, tricking creatures of the deep into thinking they're in hundreds of feet of water when they're actually in 50 feet (15 m). It's an instant deep dive for divers willing to descend through the cool, disorienting cloud of freshwater to reach the warmer, clearer saltwater below.

The reward? Black coral. Red coral. Stony coral. Snake stars, glass sponges, and sea pens. They're all here, along with 160 species of warm and cold-water fish.

Crayfish (Kiwi-speak for rock lobster) in large sizes and quantities cover crevasses. Bottlenose dolphins and New Zealand fur seals frequently buzz past, while the purple-and-white *Jason mirabilis* nudibranchs and tube anemones cling to rock walls. Black coral trees, which look frosted white underwater, stretch 15 feet (4.6 m) high. These delicate formations are usually found in water way beyond recreational limits, 300 feet (91.4 m) or more, but in Fiordland you can see them in 30 feet (9 m) of water.

Bordered by 10 marine reserves, Fiordland offers easy diving—there's no current, but instead gentle slopes and walls, and an ocean floor far out of sight. Just the cold and lack of visibility at the surface require some experience.

Milford Sound (accessible by road) and Doubtful Sound (accessed by boat or over land) are two of the best spots to dive in Fiordland.

Jacques Cousteau called Fiordland the last frontier of underwater exploration—it's not difficult to see why. ❍

"This is a storybook kind of place. A shadowy place, with ever-changing light. Around every corner you feel like you'll make a great discovery. You feel very alone . . . With the dark water you get the emergence of deep water creatures coming up to diving depths."

—BRIAN SKERRY

An 11-armed sea star sits in a garden of strawberry sea cucumbers.

A New Zealand fur seal pokes his head through the kelp.

A triplefin fish lies among invertebrates.

MEXICO

EL BOILER

An abundance of pelagics on a remote island chain

AVERAGE WATER TEMP: 76°F (24.4°C) **AVERAGE VISIBILITY:** 67 feet (20.4 m)
AVERAGE DEPTH: 83 feet (25.3 m) **TYPE OF DIVE:** Open water

The roiling surf immediately clarifies where El Boiler got its name. Located off the northwest coast of the Revillagigedo Archipelago's San Benedicto Island, a wholly Martian-looking landscape, El Boiler might look a fright, but 100 feet (30.5 m) under the surface, the beast turns into a beauty.

Divers descend near an aesthetic pinnacle, which rises 165 feet (50.3 m) to 20 feet (6 m) below the surface. The rocky, volcanic landscape is clothed in a few hard corals, but that's not the main attraction: Revillagigedo is renowned for its animal-centric dives, and this one (once you get past the surge, current, and 24-hour boat trip to get here) is a relatively simple one. Find a quiet spot around the 100-foot (30 m) mark, hunker down, and watch the show.

Manta rays—enormous, graceful, and playful—swoop in to be cleaned by colorful orange-and-electric-blue clarion angelfish, which are endemic to the area. Usually four or five manta rays make an appearance on every dive, and they seem oddly attracted to scuba divers, swimming near for a closer look, aiming for the bubble stream. One theory is that, because this is their cleaning station, the manta rays consider scuba divers to be another perk in their spa, enjoying the feel of bubbles rippling across their bellies.

"The life we observed at Revillagigedo was beyond our imaginations. This is the wildest place in tropical North America. It is a rare place of large fish, with biodiversity as if the Galápagos Islands had opened a branch in Mexico."

—ENRIC SALA, MARINE ECOLOGIST AND AUTHOR OF *PRISTINE SEAS*

A large volcano-formed rock offers opportunity for divers and fish alike.

What You'll See: Manta Rays ● Tuna ● Galápagos, Silky, Tiger, and Silvertip Sharks ● Hammerheads ● Turtles ● Angelfish ● Clarion Angelfish ● Lobsters ● Octopuses ● Jacks ● Humpback Whales (on occasion) ● Whale Sharks

Whatever the reason, the 20-foot (6 m) manta rays get up close and personal here, making El Boiler a special dive spot. If you can bear to tear your eyes away from the winged beauties, you'll likely see sharks (silky and silvertip), tuna, lobsters, octopuses, jacks, and reef fish. And if you're very lucky, this underwater spectacle might play out under the song of a humpback whale, with the whale itself putting in its own appearance (if you're very, *very* lucky).

The Revillagigedo Archipelago, known as the Galápagos of Mexico, and also frequently referred to as Socorro, the name of the largest of its four islands, lies 336 miles (540.7 km) south of the Baja Peninsula. A submerged mountain range, the four islands are the peaks of underwater volcanoes. It's known for having the largest collection of marine megafauna in North America, but—due to its remote location—that made it a target for industrial and sport fishing.

Travel Tip:

Reward reliable operators with your business, which is essential to a successful future for the park. Follow all regulations, including the policy to avoid touching or riding manta rays. As always, allow the wildlife to dictate their preferred level of interaction. Surface conditions are usually best between November and May.

A giant manta ray puts on a show for its visitors.

The stunning neon pattern of a giant hawkfish

A quartet of resting whitetip reef sharks

Lately, though, this unique spot is garnering attention for all the right reasons. It was recently named a UNESCO World Heritage site, and in 2017 the Mexican president created Revillagigedo National Park, extending the six-nautical-mile biosphere reserve to a 57,143-square-mile (148,000 sq km) national park, the largest fully protected marine reserve in North America. The protection is a great victory for the underwater world. ❍

"If you have one chance to experience an authentic, off-shore, pristine, and well-protected marine biosphere, Revillagigedo should be it. The mantas around Revillagigedo are really friendly."

—ANDREA MARSHALL, MANTA RAY SCIENTIST AND CONSERVATIONIST

AUSTRALIA

S.S. *YONGALA*

A mysterious wreck with big marine life

AVERAGE WATER TEMP: 80°F (26.7°C) **AVERAGE VISIBILITY:** 50 feet (15 m)
AVERAGE DEPTH: 46 to 92 feet (14 to 28 m) **TYPE OF DIVE:** Wreck

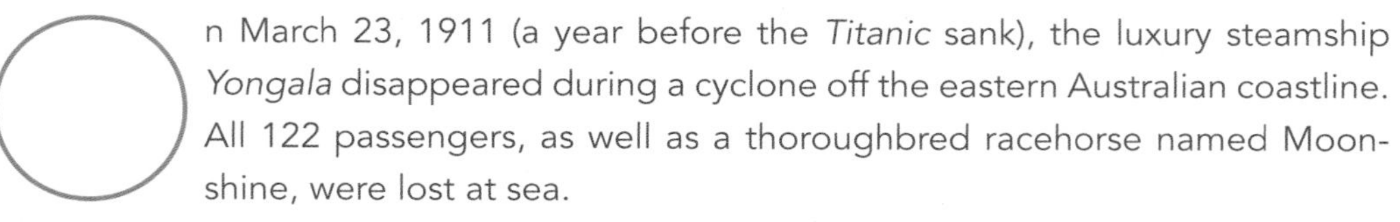

On March 23, 1911 (a year before the *Titanic* sank), the luxury steamship *Yongala* disappeared during a cyclone off the eastern Australian coastline. All 122 passengers, as well as a thoroughbred racehorse named Moonshine, were lost at sea.

Although a search was mounted, there was little to show of the *Yongala*'s fate. No life rafts or other immediate debris were discovered, although a hessian bag of cutlery with the name "*Yongala*" stamped on them was recovered, and the shark-bitten remains of a horse was found at the mouth of a creek two weeks later. It wasn't until 1958, 47 years later, that the final resting place of the S.S. *Yongala* was discovered, 48 miles (77.2 km) southeast of Townsville, Australia, less than 13 miles (20.9 km) offshore.

In a plot twist familiar to *Titanic* film enthusiasts, it was a safe that revealed the ship's identity. A local fisherman located the wreck and removed a safe and other artifacts from the vessel in the hopes of identifying it. The inside of the safe yielded nothing but sludge, but a partial serial number on its outside revealed its identity—and that of the *Yongala*.

It is now the most impressive wreck in Australian waters, and a favorite spot for scuba divers, attracting more than 10,000 underwater visitors every year.

More than 350 feet (106.7 m) long, the S.S. *Yongala* lies within the Great Barrier Reef Marine Park, resting on her side at the sandy bottom with her bow pointing north. Though the ship is surprisingly intact, it is illegal to enter the wreck—it's protected as a historical site and as a graveyard.

Still, there's plenty to see around the sunken craft. The *Yongala* is also protected as part of the Great Barrier Reef Marine Park (no fishing, no collecting), and it is also the only hard structure on a sandy seabed, washed by rivers of nutrient-rich current.

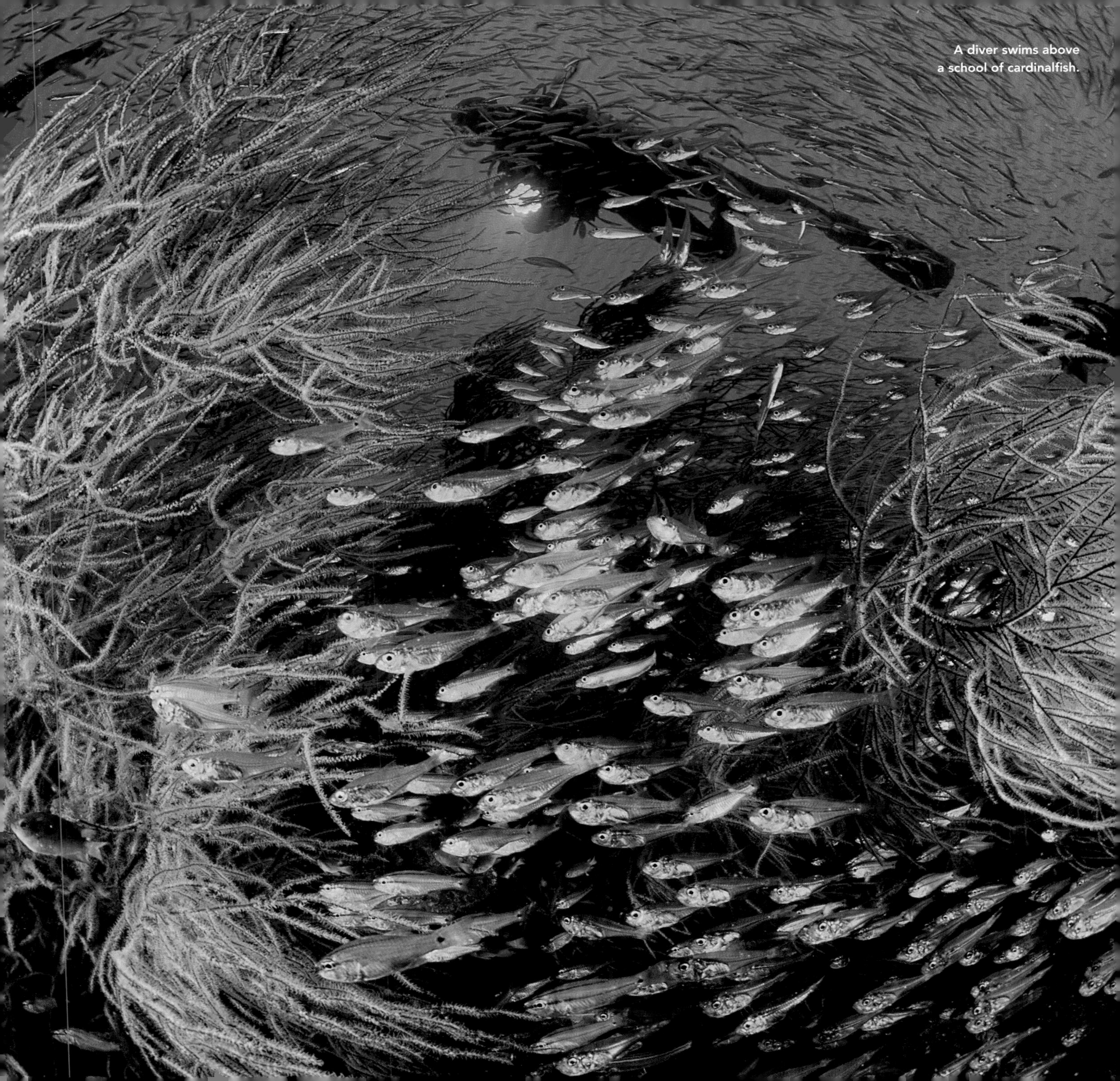

A diver swims above a school of cardinalfish.

Ocean life thrives around the wreckage of the S.S. *Yongala.*

What You'll See: Queensland Groupers • Banded Sea Snakes • Chevron Barracuda • Hawksbill, Loggerhead, and Green Turtles • Bull Sharks • Humpback Whales • Spotted Eagle Rays • Napoleon Wrasses • Giant Trevallies • Guitarfish • Moray Eels • Marble Rays

The result? Every inch of this massive ship teems with life. Soft corals cling to the hull, swarmed with tiny tropical fish. Barracuda and painted sweetlips make regular patrols along the top of the wreck, often sweeping around divers during safety stops. Banded sea snakes make a patient trek between the wreck to feed and the surface to breathe, while marble and eagle rays cruise past, splitting divers' attention between the ship and wildlife.

The turtles (loggerhead, hawksbill, and green) are so large they seem to have emerged from the age of the dinosaurs, while a local Queensland grouper has garnered the name "VW" for being as large as his namesake car.

Strong currents and its open water location mean that divers need to have their wits about them, but the rewards are worth the effort: a spectacular wreck dive featuring a staggering abundance of marine life. This is a location worthy of multiple dives, which is why many divers opt for a liveaboard, although day trips are available. ❍

Travel Tip:

Feeling brave? A night dive on the *Yongala* is an extraordinary experience, bringing in the night shift marine life—sharks, hawksbill turtles, bull rays, and eagle rays.

AUSTRALIA

LORD HOWE ISLAND

The southernmost coral reef in the world has its own unique ecosystem.

AVERAGE WATER TEMP: 68°F (20°C) **AVERAGE VISIBILITY:** 98 feet (29.9 m)
AVERAGE DEPTH: 26 to 92 feet (7.9 to 28 m) **TYPE OF DIVE:** Open water and reef

Lord Howe Island is only a two-hour flight from Sydney and is postcard-pretty. Yet this crescent-shaped UNESCO World Heritage site and marine park isn't flush with visitors. Perhaps that's because tourist numbers are restricted to 400 on the island at any given time (equivalent to the number of full-time residents). Perhaps it's because diving is still relatively new here. Whatever the reason, divers are thankful, because diving Lord Howe is like exploring a mythical planet.

For millions of years, the island was left untouched. When explorers first arrived in the late 18th century, their presence almost upset the ecological apple cart. The pigs, goats, cats, and rats they introduced to Lord Howe promptly caused the extinction of at least five species of birds.

Balance is being restored: The number of settlers and visitors is limited. Pigs and cats were removed. They're working on the rats and mice. And with no rivers draining into the lagoon, no nearby commercial fishing, and a young history of diving, the coral reefs remain some of the most pristine in the world, boasting a unique mix of tropical, subtropical, and temperate species, including many endemic (like the McCulloch's anemonefish and the Lord Howe moray eel) and rare species (like Japanese boarfish, green jobfish, and black coral).

With more than 50 locations to choose from, including reefs, caves, drop-offs, and

"Five ocean currents collide here creating an arc of oceanic life in the shadow of two towering peaks that are the remnants of an ancient volcano. My most favorite character here is the endemic double-headed wrasse."

—DAVID DOUBILET, UNDERWATER PHOTOGRAPHER

An aerial view of
Lord Howe Island

At The Archway site, a lone black spot goatfish swims among dotted sweetlips.

What You'll See: Galápagos Sharks • Japanese Boarfish • Butterfly Cod • Amberjacks • Surgeonfish • Black Coral • Spanish Dancers • Feather Stars • Lord Howe Moray Eels • McCulloch's Anemonefish

trenches, divers are spoiled. Comets Hole (a freshwater-formed hole 26 feet/7.9 m deep) is home to McCulloch's anemonefish, as well as marble shrimp, lionfish, and the Lord Howe moray eel. Tenth of June is a pinnacle and magnet for rare species, including the green jobfish and Japanese boarfish.

But it's Ball's Pyramid—the world's largest black basalt sea stack at an imposing 1,800 feet (548.6 m) —that is Lord Howe's premier site, and one that requires advanced certification due to current. This spot features drift dives, wall dives, and a 150-foot-deep (45.7 m) cave. In surprisingly clear visibility, divers can spot rare deep-water ballina angelfish, dolphins, trevallies, kingfish, and even the occasional dolphins, whale sharks, or marlins. Galápagos sharks like to crowd divers during their safety stops.

Any amount of time exploring the rare world of Lord Howe won't be enough, but it's better to be one of the 400 on the island than to miss out on seeing this gem altogether. ❍

Travel Tip:

With 400 spots for tourists on the island, book well in advance, and be mindful of the strict weight limits on flights—visitors are restricted to one 30-pound (13.6-kg) bag and one 19-pound (8.6-kg) bag each. (Rent your dive gear on the island. It's easier.) Diveable all year round, September through June is the best season to hit the waters; winds pick up in July and August.

MICRONESIA

CHUUK (TRUK) LAGOON

The ghost fleet of World War II

AVERAGE WATER TEMP: 84°F (28.9°C) **AVERAGE VISIBILITY:** 50 feet (15 m)
AVERAGE DEPTH: 75 feet (22.9 m) **TYPE OF DIVE:** Wreck

Lying under the calm waters of Chuuk (also known as Truk) Lagoon is a 40-mile-wide (64.4 km) expanse that is now the graveyard to some 60 ships and 200 planes. Nowhere else in the world do so many wrecks lie in such proximity to each other.

Located 600 miles (965.6 km) southeast of Guam, Chuuk is one of the four main island states of the Federated States of Micronesia. Its mountainous terrain, fringing barrier reef (stretching 140 miles/225.3 km), and deep lagoon caught the eye of the Japanese, who used it as a stronghold in WWII. More than 1,000 ships and 500 aircraft were stationed here at one point. The island's low coral and remarkably sheltered reef made it a perfect spot for a navy to reside.

In February 1944, the United States launched Operation Hailstone—often referred to as Japan's Pearl Harbor—a surprise naval air attack that lasted three days. It decimated the Japanese fleet—more than 230 aircraft were destroyed and more than 50

The remains of the *Fujikawa Maru*'s deck

Amid the wreckage, marine wildlife thrives.

CALIFORNIA

MONTEREY BAY

Delight in giant kelp forests and abundant marine life.

AVERAGE WATER TEMP: 50°F (10°C) **AVERAGE VISIBILITY:** 23 feet (7 m)
AVERAGE DEPTH: 60 feet (18 m) **TYPE OF DIVE:** Open water and shore

California's Monterey Bay is synonymous with the ocean, from its world-class aquarium to its luxurious kelp forests. This spot has attracted divers for decades, with deep water close to shore and a rich variety of marine life.

Located on the lip of the Monterey Bay Submarine Canyon, an underwater rift rivaling the Grand Canyon in scale, Monterey is blessed with nutrient-rich waters that feed both the large (giant Pacific octopuses and bluntnose sixgill sharks) and the small (Norris's and jeweled top snails and nudibranchs). Keep your underwater eyes peeled—there's no shortage of things to see.

Shore dives—like the Breakwater, a man-made site accessible from a U.S. Coast Guard pier, or San Carlos Beach, with its population of playful sea lions—are popular. Only a five-minute swim over a sandy/silty bottom leads to an ocean floor carpeted with colorful sea stars, anemones, and sea urchins that attract sea otters hunting for a snack.

A short boat ride opens up even more opportunities. Hopkins Deep Reef (70 feet/21 m) is a great location to spot nudibranchs (including Spanish shawls), hermit crabs, and cancer crabs. Take a close look among the anemones for octopuses. Outer Chase Reef is a collection of pinnacles, home to bright anemones and sea stars.

If you need any advice, the ocean-loving locals are the ones to ask. Although the area attracts more than 65,000 divers annually from around the world, it's a local favorite, well loved and well protected, a true backyard dive hangout. Monterey divers need cold-water experience, with at least a 7mm wet suit or a dry suit. ○

What You'll See: Kelp Forests • Zebra and Bluntnose Sixgill Sharks • Sea Otters • Sea Lions • Giant Pacific Octopuses • Norris's Jeweled Top Snails • Nudibranchs • Sea Stars • Sea Urchins • Anemones • Pacific Sea Nettles • Wolf Eels

A jellyfish swims through lush kelp forests.

COSTA RICA

COCOS ISLAND

Where pura vida *means a treasure trove of marine life*

AVERAGE WATER TEMP: 81°F (27°C) **AVERAGE VISIBILITY:** 70 feet (21 m)
AVERAGE DEPTH: 80 feet (24 m) **TYPE OF DIVE:** Open water, reef, and wall

Historians estimate there's more than a billion dollars of plundered pirate treasure buried on Cocos Island.

It's a good hiding spot. Located 342 miles (550.4 km) off Costa Rica in the vast Pacific Ocean (and almost the same distance from the Galápagos Islands), Cocos is the only eastern Pacific island with a tropical rain forest, a green canopy of camouflage for a small emerald piece of land, a mere five miles (8 km) long, easy to miss.

But if you're thinking about hunting for some ill-gotten gold to fund future diving trips, think again: Cocos Island is now a protected UNESCO World Heritage site.

Don't be too heartbroken. Cocos's real treasure lies under the surface: Schooling hammerhead sharks, 100 or more, circle in the converging, nutrient-rich currents of the Golden Triangle, an area extending from Malpelo Island (page 350) to the northernmost Galápagos Islands (Wolf and Darwin, page 270) to Cocos Island. This area is home to some of the most impressive and abundant marine life found in the ocean, and you can share the water with them in Cocos. Manta rays cruise past looking for

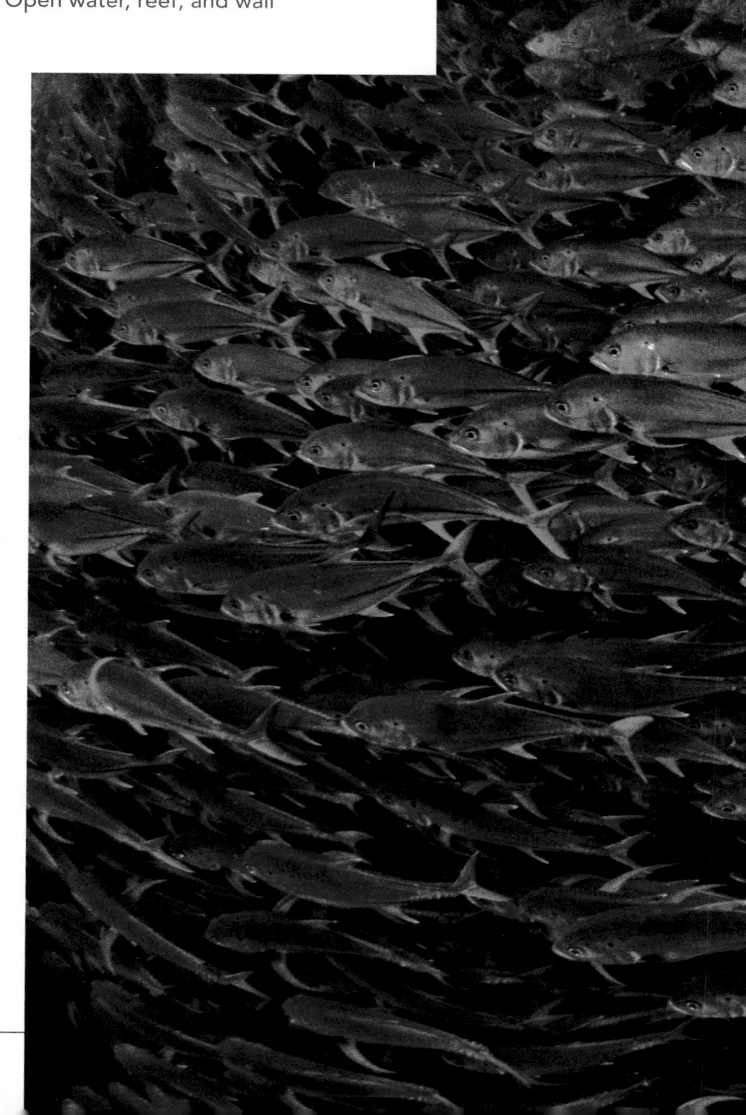

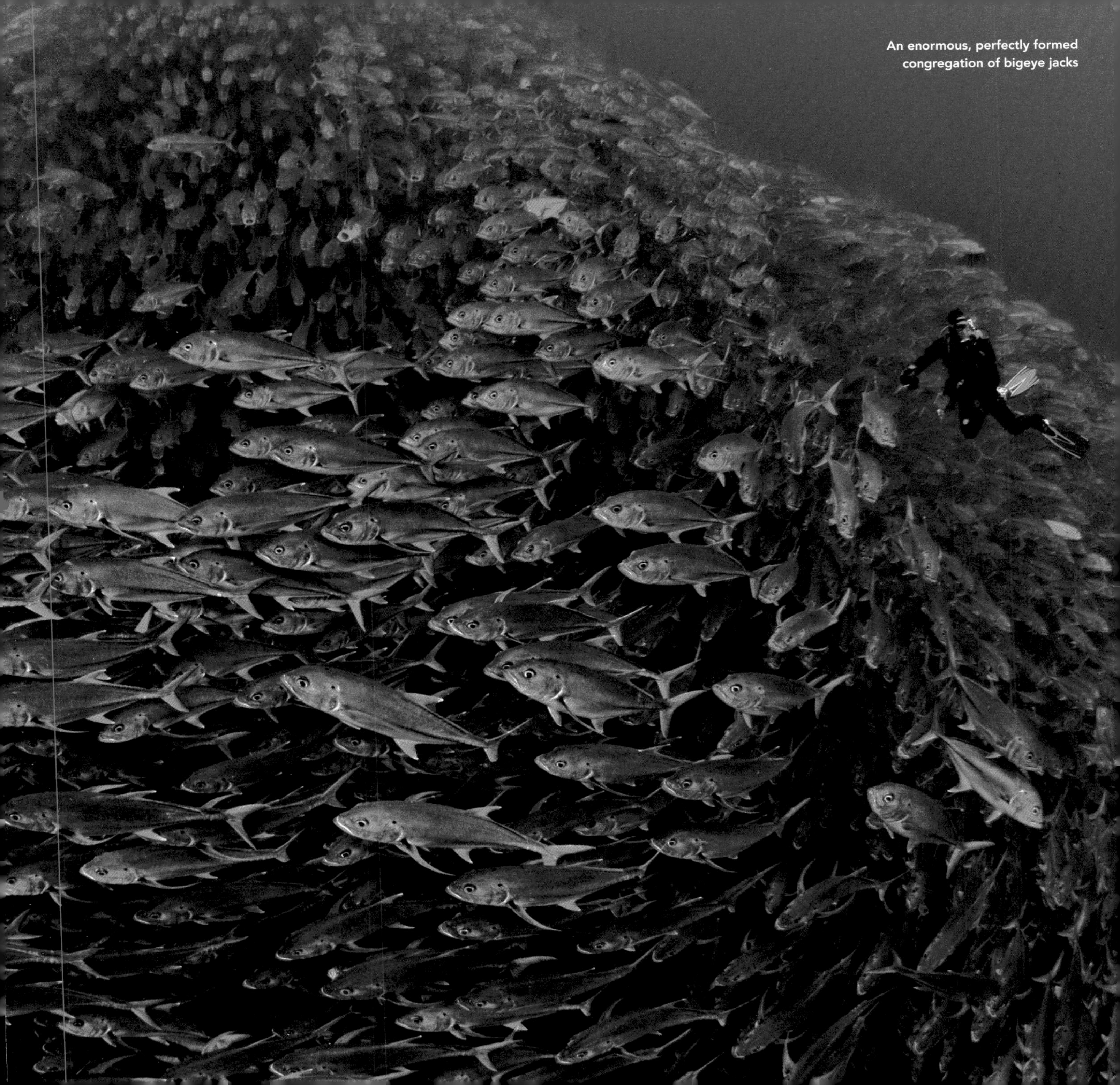

An enormous, perfectly formed congregation of bigeye jacks

What You'll See: Scalloped Hammerheads • Tiger, Blacktip Reef, Silky, Galápagos, and Whitetip Reef Sharks • Swordfish • Marlin • Bottlenose Dolphins • Whale Sharks • Yellowfin Tuna • Horse-Eye Jacks • Marbled, Mobula, and Eagle Rays • Manta Rays • Moray Eels • Green and Hawksbill Turtles

cleaning stations, and other sharks (tiger, blacktip reef, silky, Galápagos, whitetip reef) and rays (marbled, mobula, eagle) regularly make appearances. Gems of the ocean, everywhere you look. Accessed by liveaboard only (and a nonstop 36-hour crossing), Cocos is home to 20 unique dive spots, ranging from shallow reef dives to vertical walls to pinnacles.

Bajo Alcyone is hammerhead central. Located about a mile (1.6 km) offshore, divers descend 82 feet (25 m) to the seamount top, seeking out a sheltered rock space to watch hammer time. These graceful creatures prefer cruising below the thermocline, luring many a diver out of their hiding place to pursue them below 100 feet (30.5 m). (Keep an eye on your depth—that, paired with the strong current, can make this spot a tricky one.) Cleaning stations service silky and Galápagos sharks, as well as manta rays, and mind which rock you grab on to—an octopus might have its eye on the same one.

Dirty Rock is the other hammerhead showcase. Divers descend steadily to 75 feet (22.9 m), sit back, and watch as hammerhead sharks, marble rays, goatfish, bluefin trevallies, and Amarillo snappers crowd in, sometimes unsettlingly close. This site is a mix of rocky pinnacles and boulders, separated by a deep (300-plus feet/91.4+ m) channel, which is why it's a popular spot for large and plentiful fish. You might even see a whale shark slide through, and bottlenose dolphins have been known to dart around divers during safety stops. December through May brings calmer seas and better visibility, but the rainy season—June through December—brings in more hammerhead sharks, whale sharks, and manta rays.

If you need a calming cruise after all that large fish action, Manuelita Garden, a coral garden on Cocos's protected east side, is just the place. This is one of the few coral reef dives at Cocos, and it is home to spiny lobster, parrotfish, damselfish, and goatfish that delight divers. Make sure you take the occasional look over your shoulder, as tiger sharks regularly visit this area.

From any viewpoint—the underwater wonderland to the waterfall-laden, lush green rain forest—Cocos lives up to its rich reputation. If weather and conditions permit, liveaboard guests may have the opportunity to explore Cocos on foot, an extraordinary one-of-a-kind experience. ❍

The uninhabited Cocos Island

A moray eel comes out from its hiding spot.

Brown boobies check underwater for lunch.

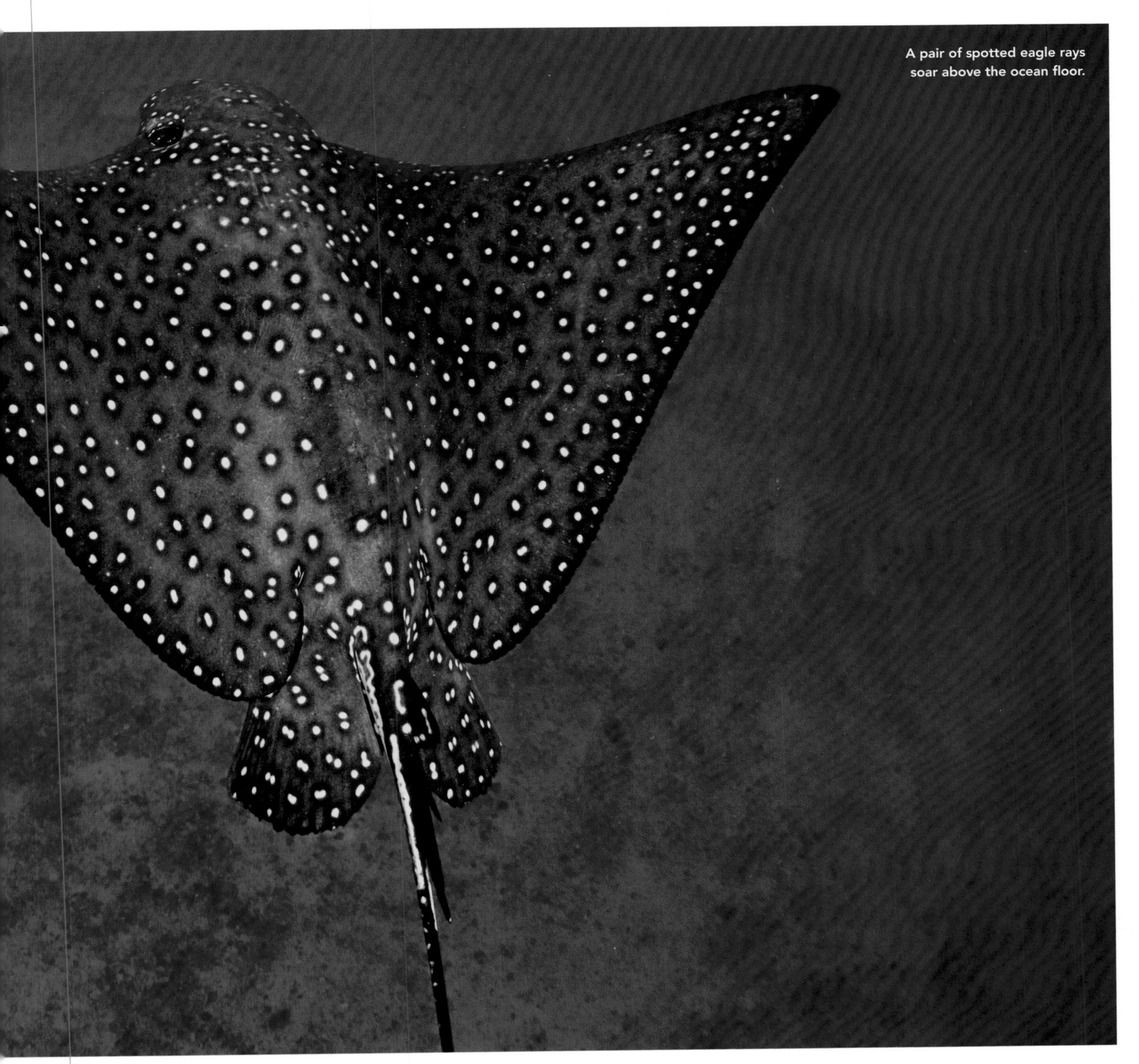
A pair of spotted eagle rays soar above the ocean floor.

BRITISH COLUMBIA

BARKLEY SOUND

One of the world's most colorful cold-water diving hot spots

AVERAGE WATER TEMP: 49°F (9.4°C) **AVERAGE VISIBILITY:** 100 feet (30 m)
AVERAGE DEPTH: 60 feet (18 m) **TYPE OF DIVE:** Cold water

Barkley Sound is a burst of cold-water color. Swaying gently in an emerald sea, anemones, nudibranchs, and fish as brightly hued as any tropical reef fish light up the underwater world in oranges, lilacs, pinks, and yellows. Here, there's something new to see on every dive, with plenty of sites to cope with whatever the British Columbia weather is up to. This is why it's a cold-water hot spot.

Located south of Ucluelet and north of Bamfield on Vancouver Island's west coast, this 309-square-mile (800.3 sq km) area is home to hundreds of small islands.

With rich marine life (kelp forests, wolf eels, and visiting orcas), wrecks (at least 60 in the area, earning it the nickname "graveyard of the Pacific"), and reefs, Barkley Sound is one of those places you'd need a lifetime to fully explore.

Renate Reef is one of the Sound's premier spots, a pinnacle with a flat top located in the Imperial Eagle Channel. Renate is a mix of invertebrates and fish, with lingcod, strawberry anemones, rockfish, and giant Pacific octopuses calling it home.

Mahk is another reef—several parallel reefs, actually—home to ratfish, which are cuter than the name implies. During the summer months, they seek out shallower waters.

"Between the Broken Islands there are an astonishing number of wild things above and below the water. The fjord ecosystem of Effingham Inlet creates anoxic zones where life thrives, just not the life we're used to. The scale of the oceans is enormous, but the scale of life is even more astounding."

—ERIKA BERGMAN, SUBMARINE PILOT AND NATIONAL GEOGRAPHIC EXPLORER

A topside view of Tzartus Island
on the Barkley Sound

Sea stars, kelp, and coral make up just part of the Sound's abundant sea life.

What You'll See: Sea Cucumbers • Anemones • Wolf Eels • Nudibranchs • Kelp Forests • Giant Pacific Octopuses • Sixgill Sharks • Sea Pens • Orcas • Puget Sound King Crab • Sea Snails

Another spot for unique marine life is Tyler Rock, renowned for sixgill shark sightings during June and October. Giant Pacific octopuses, strawberry anemones, rockfish, and wolf eels are also regularly found here, and the occasional humpback whale is known to pass through town.

Kyen Point, another Imperial Eagle Channel spot, is comprised of two pinnacles with kelp beds swaying in the channel that runs between them. Along with wolf eels and octopuses, divers might catch sight of playful sea lions. Another point, Chup, is a macrophotographer's dream: Sea pens, nudibranchs, scallops, anemones, and purple-ringed top snails carpet the underwater landscape in vivid color.

And then there's the *Vanlene*. After encountering thick fog, the freighter ran into Austin Island, shattering its bow. It took several days to sink, so most of its cargo of Dodge Colt cars (it was 1972) was safely off-loaded. The stern is largely intact, and it's now home to schools of rockfish and Puget Sound king crab. ❍

Travel Tip:

Barkley Sound can be dived year-round. Even in winter, with snow and temperatures dropping to 12°F (−11.1°C), the geographic makeup of Barkley shelters divers from any winter storms. Algae blooms can reduce visibility to 30 feet (9 m) in the summer, but it usually improves at depth.

FIJI

GREAT ASTROLABE REEF

Dive a wild reef with channels, pinnacles, and vibrant coral.

AVERAGE WATER TEMP: 81°F (27°C) **AVERAGE VISIBILITY:** 80 feet (24 m)
AVERAGE DEPTH: 15 to 110 feet (4.6 to 33.5 m) **TYPE OF DIVE:** Open water and reef

Kadavu is pure Fiji. Growing in infrastructure, it still retains its lush, wild feel and strong cultural roots. Fiji's fourth largest island (159 square miles/411.8 sq km) lies smack on the fourth largest barrier reef in the world, the Great Astrolabe Reef, which stretches more than 60 miles (96.6 km).

The Astrolabe's coral growth is remarkable: finger, table, bottlebrush, and staghorn corals, as well as sponges, sea whips, and sea fans, in vibrant violets and greens clash with yellow sunset wrasses, hawkfish, and butterflyfish. Passages in the reef funnel strong currents favored by reef sharks, groupers, and manta rays, which frequent cleaning stations at the southern part of Kadavu.

A short boat ride is all it takes to reach most dive sites, like Solo Reef, a section teeming with fish that will approach patient divers.

Eagle Rock is a Great Astrolabe showcase, with swift passages, shark pinnacles, and hard corals, while Sea Fan Alley on Kadavu's western end is a slope leading to a deep blue drop-off. Yellow and green sea fans cling to the reef, circled by angelfish, triggerfish, and sea snakes. Hammerheads can sometimes be spotted here, cruising the indigo depths.

Kadavu's untamed nature extends to its underwater world—there's still an exploratory feel to diving here, with currents and swells mixing with sheltered bays. Don your best Jacques Cousteau mindset and set off to look around a reef that is still wild. ○

What You'll See: Groupers • Manta Rays • Butterflyfish • Hawkfish • Finger, Staghorn, and Table Coral • Sea Whips • Sea Fans • Snapper • Sweetlips • Napoleon Wrasses • Giant Clams • Green Turtles • Humphead Wrasses • Potato Cod

Dusky anemonefish swim over
a large colony of sea anemones.

VANUATU

S.S. *PRESIDENT COOLIDGE*

Visit the Lady, 125 feet (38 m) within Vanuatu's most famous wreck.

AVERAGE WATER TEMP: 84°F (28.9°C) **AVERAGE VISIBILITY:** 65 feet (19.8 m)
AVERAGE DEPTH: 100 feet (30 m) **TYPE OF DIVE:** Wreck

The S.S. *President Coolidge* has lived many lives. She began as a luxury passenger steamship, built in Virginia in 1931. No expense was spared on this 615-foot-long (187.5 m) floating art deco hotel, one that regularly broke speed records. It came complete with living space for nearly 1,000 passengers, gleaming chandeliers, a gym, a doctor's office, stock exchange, barber shop, and saltwater swimming pools.

When WWII broke out, the S.S. *Coolidge* rolled up her sleeves and pitched in, evacuating Americans from Hong Kong and other parts of Asia, and transporting Pearl Harbor's wounded to San Francisco.

As the war intensified, the S.S. *Coolidge* was transformed: Her art deco interior was painted gray, the passenger space was redesigned to accommodate 5,000 troops, and she was stacked with antiaircraft guns.

In 1942, within sight of Espiritu Santo in Vanuatu (see Million Dollar Point, page 138), the S.S. *Coolidge* approached, carrying more than 5,000 army and navy troops and tons of cargo, everything from howitzers to jeeps to ammunition to 500 pounds (227 kg) of quinine. Only one thing was forgotten—the vital information that the Segond Channel, where the S.S. *Coolidge* was headed, had recently been mined.

Two explosions tore into the ship in rapid succession, and then she hit a coral reef, fell on her port side, and slipped into deeper water. Luckily, all but two men from the ship's 5,340 troops and crew had time to make it off the ship. Now, the S.S. *Coolidge* comprises a week's worth of dive sites, easily accessible by shore or boat. Most are decompression dives, as the highest point of the ship lies approximately 68 feet (20.7 m)

The famous Lady of the S.S. *Coolidge*

Divers explore the S.S. *Coolidge*'s forward guntub.

What You'll See: Triggerfish • Lionfish • Moray Eels • Nudibranchs • Barracuda • Coral • Sponges • Anemones • Groupers • Angelfish • Leaf Fish • Ghost Pipefish

below the surface, stretching to more than 200 feet (61 m) long. Although earthquakes have inflicted some damage on the ship, it's mostly intact and slowly turning into an artificial reef, providing a home for all manner of marine life.

Those with the proper dive training have a huge shipwreck to explore. Swim the saltwater swimming pools. Visit a cargo hold with Willys jeeps and howitzers. Drop 108 feet (32.9 m) to explore the stash of medical supplies on decks B, C, and D and other relics of war.

One of the most sought-after sights is the Lady, a porcelain relief panel of a woman riding a unicorn, tucked away 125 feet (38 m) inside the first-class dining room. She has to be one of the most popular underwater photograph subjects in the world.

And for experienced divers, the stern is a siren call. A massive rudder lies in 200 feet (61 m) of water (nitrogen narcosis is a very real risk here), the brass S.S. *Coolidge* lettering on the back of the ship almost a foot high, with a return swim that takes you nearly the entire length of the ship. ❍

Travel Tip:

Plan your dive and dive your plan, particularly with the S.S. *Coolidge*. Most of these dives have decompression times, but there's so much to see that bottom time and depth can easily sneak away from you here. Keep your wits about you.

ERITREA

DAHLAK ARCHIPELAGO

The Red Sea without the crowds

AVERAGE WATER TEMP: 84°F (28.9°C) **AVERAGE VISIBILITY:** 27 feet (8.2 m)
AVERAGE DEPTH: 73 feet (22.3 m) **TYPE OF DIVE:** Open water and lagoon

If you're looking for a dive that's off the beaten track, your compass should point to the Dahlak Archipelago in Eritrea.

Located 30 miles (48.3 km) out from Massawa, on the western side of the Red Sea across from Saudi Arabia's Farasan Islands, the Dahlak Archipelago is a scattered collection of two main islands and more than 120 smaller islands, only four of which are inhabited.

It's a diving dichotomy. On the one hand, Eritrea, an impoverished nation caught between a rock and a hard place (it's bordered by Sudan, Ethiopia, Djibouti, and the Red Sea) has earned itself the unfortunate nickname of the "North Korea of Africa." Decades of repression and forced military conscription under leaders accused of crimes against humanity by the United Nations has led to unrest, exodus, and a lackluster tourism industry.

On the flip side, the lack of tourists has inadvertently led to a reprieve for sites that would normally have been overvisited by Red Sea crowds. War shut down the fishing industry, resulting in a staggering increase in fish populations. And a group of Eritrean naval freedom fighters have found new employment in the fluttering dive tourism industry.

There are reportedly around 15 (uncrowded) dive sites. Norah, a shallow lagoon, teems with reef fish such as parrotfish, clownfish, and butterflyfish, along with bright corals, making it a good spot for underwater photographers.

For larger life like groupers, barracuda, and reef sharks, the underwater rock formations of Northwest, Dessei Islands, and Sehil Camel should be your port of call.

Dahlak Kebir is the largest in the archipelago, with interesting underwater explorations (including a coral-crowded Russian dry dock, with schools of fish cruising around old lamps

A pair of clownfish watch over eggs.

A bicolor parrotfish seems to smile for the camera.

What You'll See: Hermit Crabs • Butterflyfish • Parrotfish • Red Snappers • Pufferfish • Reef Sharks • Moray Eels • Clownfish

and other spare parts), as well as aboveground cultural and historical points (ancient Islamic and Turkish ruins, including cisterns and wells). The nearby marine base of Nokra, set up by the U.S.S.R., is now a home for coral and reef fish, as well as manta rays.

The Dahlak Archipelago is rare, remote, unvisited, and a place where divers can feel miles away from civilization. But getting underwater isn't easy. At the moment, permits are required, and divers can only go with licensed boats (no fishing boats or charters). Ordinarily this wouldn't be an issue, except that the fees being charged to dive the Dahlaks are highway robbery.

Conditions and political climates change, so if you're feeling the siren call of exploration, keep a watchful eye on this destination.

Marine biologists are doing the same. The shallow, warmer seas lapping against Eritrea's shores seem to have created corals that can adapt to increased temperatures. Could this be a help for coral reefs threatened by global warming? ○

Travel Tip:

If you're traveling to Eritrea's capital city of Asmara, take note that it's above sea level. Add in extra time to minimize the risk of decompression sickness. Malaria precautions are essential, but the side effects of malaria medication can pose a risk for divers. This is a destination where it's essential to do your homework.

SOUTH CAROLINA

COOPER RIVER

Hunt for the teeth of an ancient giant.

AVERAGE WATER TEMP: 67°F (19.4°C) **AVERAGE VISIBILITY:** none
AVERAGE DEPTH: 30 feet (9 m) **TYPE OF DIVE:** Muck

Chilly. Fast-flowing. Zero visibility. And alligators. If those are the conditions the Cooper River has to offer, why on Earth is it a diving hot spot?

The answer—megalodon teeth. The Cooper River is full of them.

Megalodon (which are definitely extinct, by the way) were prehistoric sharks from the Miocene epoch that dwarf present-day great white sharks. One shark expert even memorably described a great white as being the size of a clasper of a male megalodon.

Growing 50 feet (15 m) in length, with six-inch (15.2 cm) teeth, megalodon had one of the most powerful bites—if not *the* most powerful bite—of any creature that has ever lived (megalodon literally means "mega tooth.") It's fitting, then, that all that remains of this behemoth are its teeth. The rest of megalodon's skeleton was mostly composed of cartilage, so what little we know of these giants are hand-size triangles scattered throughout the world, including in the Cooper River.

The gravel and silt riverbed of the Cooper River is a megalodon hot spot, as well as a fossil deposit for mammoths, saber-toothed cats, and cave bears.

Operators take brave fossil hunters out to the gravel beds, where divers descend carrying strong flashlights, extra weights (to descend quickly and stay there), nets (for carrying the spoils), something to anchor themselves against the current (screwdrivers are the weapon of choice), and a belly full of courage.

Using one hand to dig, fan, and rake through the gravel or untangle themselves from submerged trees they hope aren't alligators (although seen topside, very few divers report seeing one underwater), the payoff is at the very least a heady adventure. And who knows? It might be prehistoric gold. ❍

What You'll See: Nothing (including your dive buddy)

The sun sets over a pier along the Cooper River.

Hammerheads circle close to the water's surface.

ECUADOR

DARWIN ISLAND

An extraordinary, exotic, evolutionary cauldron

AVERAGE WATER TEMP: 73°F (22.8°C) **AVERAGE VISIBILITY:** 52 feet (15.8 m)
AVERAGE DEPTH: 65 feet (19.8 m) **TYPE OF DIVE:** Open water

If you're ho-hum about nature, the Galápagos Islands are an instant cure. This living laboratory has the primeval feel of time accelerating and standing still all at once. This is an archipelago where nearly 95 percent of the prehuman biodiversity remains intact, and evolution can be observed occurring in as little as three generations. Everything that lives on, above, or under these volcanic islands flew, swam, or adapted here. It's an extraordinary place with exotic endemics—Galápagos penguins (the only species found north of the Equator), iguanas, and giant tortoises.

Located 620 miles (997.8 km) from the South American mainland, the convergence point for three ocean currents, the Galápagos Islands were known to whalers, sailors, sealers, and buccaneers,

What You'll See: Galápagos and Silky Sharks • Hammerheads • Whale Sharks • Manta Rays • Galápagos Penguins • Iguanas • Moray Eels • Marbled and Spotted Eagle Rays • Creole Fish • King Angelfish • Green Turtles • Galápagos Sea Lions • Bigeye Jacks • Lobsters

Confronting a diver, a whale shark shows its impressive size.

but it was the H.M.S. *Beagle's* 1835 visit that changed everything. Notebook-toting Charles Darwin observed that, although all of the Galápagos Islands shared similar climate, environment, and volcanic composition, each island was home to its own unique species. That observance was integral to Darwin's theory of evolution.

"Charles Darwin made the Galápagos Islands famous, but for the underwater world to be so full of life is something he probably never imagined," said Enric Sala, National Geographic explorer-in-residence and leader of the Society's Pristine Seas project.

Luckily, divers don't need to imagine it—you can see for yourself.

Darwin Island is the northernmost island of the Galápagos Archipelago, and the diving

"The Galapagos left me in awe of nature, more so than many places that I have traveled. The animals are so abundant and also unafraid of people. I felt very much like a visitor to their ecosystem—and a fortunate one at that. From the sea lions to the sharks, I felt that my curiosity was reciprocated and it filled me with joy, knowing that places like this still exist."

—JESS CRAMP, SHARK RESEARCHER AND MARINE CONSERVATIONIST

hot spot of the area. It isn't an easy dive: a long and bumpy boat ride, 7mm cold-water gear, strong swells and current, and average visibility mean this isn't beginner territory, but the wealth underwater is worth it.

A diver greets a Galápagos green turtle.

Travel Tip:

The government of Ecuador is trying to protect the islands by limiting the number of visitors, so keep that in mind when planning your visit, along with the fact that islands tend to be quieter December through May (barring the holiday season). Diving is by liveaboard.

Where else can you swim with schools of hammerheads, Galápagos sharks, Galápagos penguins, iguanas, whale sharks, manta rays, and Galápagos sea lions, all in the same day? (Just don't try to follow the last—they can dive 1,970 feet/600.5 m underwater, holding their breath for up to 30 minutes.)

The legendary Darwin's Arch marks the spot like an X on a map. Divers back-roll out of small boats and descend through medium current and chop underneath the silhouettes of schooling hammerheads, buzzed by eagle and manta rays, surrounded by schools of trevally, barracuda, and colorful parrotfish and raccoon butterflyfish.

Dive plans alter depending on weather conditions—from tucking yourself in away from the current and watching the show, to a rubble slope known as the Darwin Theatre, to roaming around a 65-foot-deep (19.8 m) sandy flat home to hundreds of garden eels.

Although you will see plenty of sharks (the Galápagos is one of the world's most densely populated shark zones), bear in mind that illegal fishing has already wiped out 90 percent of the world's shark populations, and even protected areas aren't safe. In 2017, a vessel was intercepted with more than 300 tons (272.2 metric tons) of endangered silky and hammerhead sharks from the Galápagos, and more ships linger on the fringes of the 51,350-square-mile (133,000 sq km) reserve.

But there is hope: That same year, scientists discovered a rare nursery for endangered scalloped hammerheads in the mangrove-lined coast of Santa Cruz Island, where live-born pups feed on crustaceans in protected areas before heading out to sea. The nursery's location has been kept a secret to protect the baby sharks from harm. Hopefully it will remain that way. (And if it doesn't, please help protect—rather than exploit—this rare find.)

If you're lucky enough to dive Darwin Island, and you fall in love with what you see, use your voice to protect it. ❍

A rare sight: The marine iguana swims underwater.

PORTUGAL

AZORES

Diving as unusual and varied as the islands themselves

AVERAGE WATER TEMP: 67°F (19.4°C) **AVERAGE VISIBILITY:** 100 feet (30 m)
AVERAGE DEPTH: 15 feet (4.6 m) to beyond recreational limits **TYPE OF DIVE:** Open water

Imagine Hawaii's verdant green landscape, fringed by blue ocean and threaded with silver waterfalls. Now add a few Italian villages.

That alluring image is the Azores, a Portuguese territory of nine small islands scattered in the North Atlantic, the tip of a triangle formed between the western coasts of Morocco and Portugal.

The Azores are the type of place travel writers prefer to keep to themselves, loathe to share the hidden treasure with prying eyes. The archipelago is considered one of the most sustainable regions in the world, with everything from the 15th-century Portuguese town of Angra do Heroismo to the vineyard culture of Pico Island, a UNESCO World Heritage site.

The good news for divers is that this world of wonders extends below the surface, an abundance of life ranging from grouper to wrasse. It's also one of the only places in the world divers can see blue sharks and mako sharks in the open ocean.

With more than 100 identified sites, there is something to please every diver, from coastal diving with enormous dusky groupers and slipper lobsters, to schools of barracuda, tuna, marlin, and jacks offshore in astounding visibility, to dramatic underwater formations—gorges, drop-offs, canyons, and arches.

The Azores are divided into three different island groupings. The eastern group consists of Santa Maria, São Miguel (home to the capital Ponta Delgada), and the Formigas Islets. Flores and Corvo make up the northwestern group. And then there's the central group, comprised of Pico, São Jorge, Terceira, Faial, and Graciosa.

Known as the "Green Island," São Miguel is the largest island in the Azores, about 54 miles (86.9 km) away from Santa Maria. It's known above water for its crater lakes and hot springs. Under the waves, the S.S. *Dori* shipwreck is a favorite spot. This WWII Liberty ship rests on sand, with the intact stern at 28 feet (8.5 m) and the bow at 60 feet (18 m).

The Gonçalo lighthouse overlooks the ocean.

A sperm whale and her calf
enjoy the azure waters.

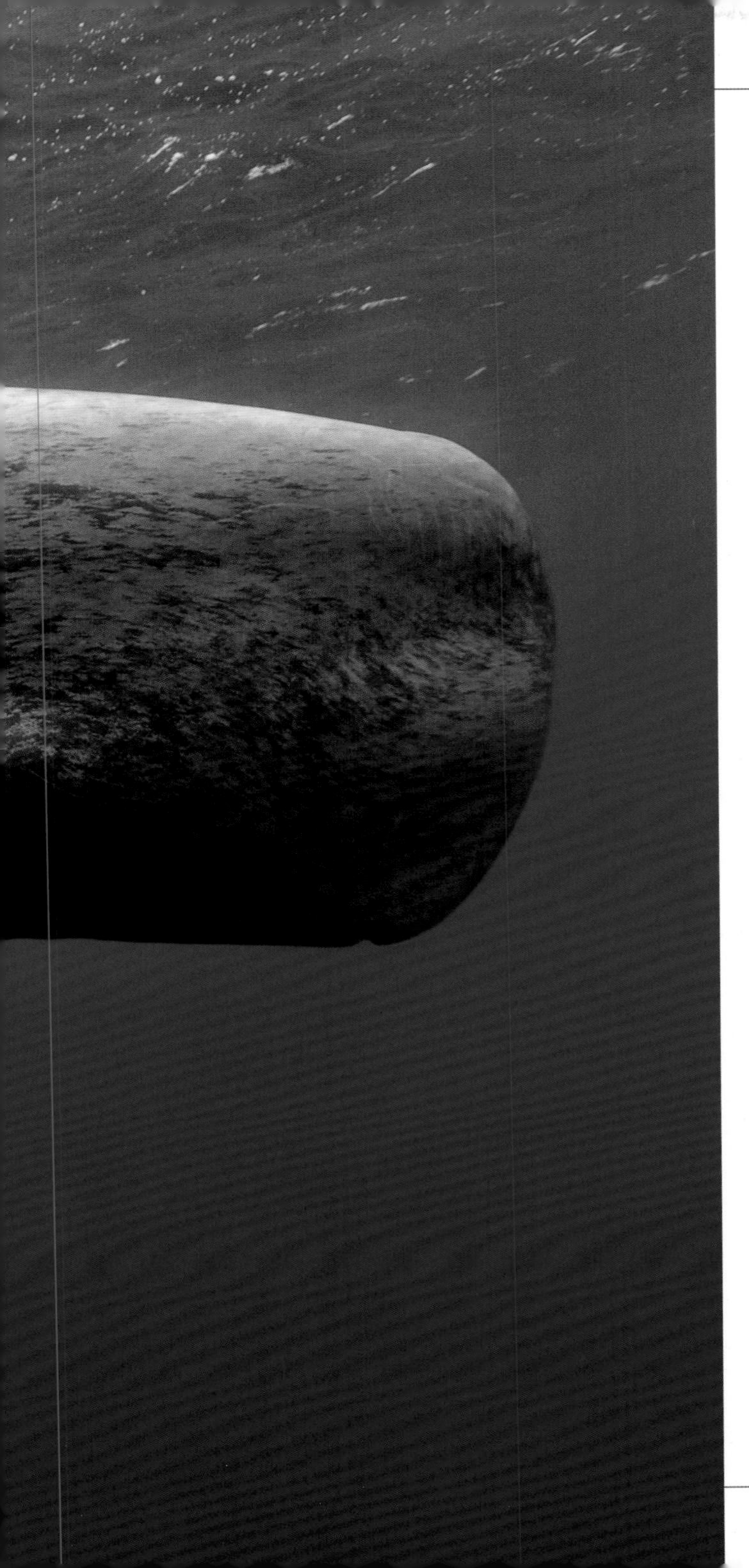

What You'll See: Blue and Mako Sharks • Mobula Rays • Dusky Groupers • Amberjacks • Bream • Wrasses • Purple Conger Eels • White Trevallies • Yellowmouth Barracuda

Santa Maria, the Island of Sun, has a protected area known as Ambrosio Reef, a mere three miles (4.8 km) from the coast. Along with rays and pelagic schools, whale sharks like to pop by in the summer.

The submerged seamounts at the Formigas Islets (approximately 40 miles/64.4 km away and accessible from both São Miguel and Santa Maria) are world-class deep-water dives, with schools of large pelagic fish, mako and blue sharks, manta rays, and mobula rays cruising clear, indigo waters.

Flores is a popular island with divers, known for its jagged coastline, caverns, and sheltered bays. Galo Cave is a cavern with two entrances where divers can surface under a fresh waterfall, while the Amigo and Morro coastal reefs are home to huge schools of white trevallies, yellowmouth barracuda, and jacks.

Nearby Corvo also has a thriving marine life, thanks in part to its isolation, and in part to the voluntary reserve established by the island. The Dusky Grouper Passageway, with its 60-pound (27.2 kg) namesakes, is a must-dive spot.

Travel Tip:

Diving usually takes place from shore-based boats in the Azores. The winter months have fewer tourists and more storms, yet the diving is surprisingly good (but cold). July to October is the high season, but a great time to spot blue sharks and manta rays.

A dusky grouper explores the reefs.

Chilean devil rays congregate in the warm waters.

So is Princess Alice Bank, roughly 50 miles (80.5 km) from Pico Island, a pinnacle reaching up within 95 feet (29 m) of the surface, swarmed by groups of 30 mobula rays, moray eels, and large pelagic life. ❍

"The Azores is one of those places where you never really know what you'll see. The region is punctuated by incredible swings in marine life from very little in the winter to hordes of life that pile into the region during summer as if someone is ringing a dinner bell. Visibility tends to be good, the water a bit on the cold side, and the megafauna is incredible."

—CAMRIN BRAUN, OCEANOGRAPHER

MALTA

THE BLUE HOLE

Descend into a sapphire blue grotto.

AVERAGE WATER TEMP: 70°F (21°C) **AVERAGE VISIBILITY:** 120 feet (36.6 m)
AVERAGE DEPTH: 49 feet (14.9 m) **TYPE OF DIVE:** Shore

It's something to say that Gozo's Blue Hole is the most beautiful gem in Malta, a medieval-looking Mediterranean island nation constructed in blinding white and mellow gold limestone, crowned by towers and domes, sunsets glinting off windows.

Malta is old. This republic perched on three islands between Sicily and Tunisia has been a prize to be won almost since war was invented. Archaeological remains of the Neolithic era, followed by the Phoenicians, Romans, and Arabs (to name a few) are evident, and its role as home base to the Order of the Knights of Malta is just downright rock star. It's not merely the history that's old, however. Archaic mindsets linger here as well—for example, divorce was illegal until 2011.

But beauty abounds everywhere you cast an eye, from the mainland to Gozo, the second largest island and a more rural landscape. With its eye on dive tourism, Gozo is already making a name for itself in underwater circles.

It's the remarkable visibility in seas hued emerald and sapphire and the arresting topography that are a winning combination for Gozo. The flaring arches, hollowed-out caverns, trenches, and caves that carve

Divers prepare to descend the Blue Hole in Malta.

The perfect photo op: Divers are silhouetted by the Blue Hole during their ascent.

the landscape into a work of art are also present underwater, and the Blue Hole is the belle of the ball, a natural rock formation made by wave and wind.

A cerulean pool framed by a rippling rock barrier wall protecting it from the ocean marks the entrance to the Blue Hole, with remnants of the famous Azure Window, an enormous arch eroded by wind and water that collapsed in 2017, holding court in the background. (It is still an impressive site.)

Divers descend 30 feet (9 m) into the blue grotto surrounded by swimmers enjoying a cool dip. An archway leads out into the open sea, a velvet environment of deep blue lit like crystal.

What You'll See: Unique Topography ● Moray Eels ● Lobster ● Octopuses ● Parrotfish ● Tuna ● Barracuda ● Groupers ● Bream

The collapse of the Azure Window has changed the underwater landscape, both creating and destroying a garden of boulders and newly formed swim-throughs, partially blocking part of the old route. Divers skirt their way around the blocks, looking for moray eels and octopuses tucked into crevices, while parrotfish, barracuda, and groupers patrol unperturbed.

After passing by the Azure Window, divers wrap up by exploring the cavern or continuing up a chimney that opens onto a coral reef.

Divers need to explore the Blue Hole with a professional guide while the landscape stabilizes itself, but that affords an excellent opportunity to utilize their local knowledge. Any time is a good time to dive the Blue Hole. The summers are hot and dry, and the winters are short and cool. Dives are usually shore dives or a short boat trip. ❍

The cliffs of Malta at dusk

BAHAMAS

TIGER BEACH

Get up close and personal with tiger sharks, lemon sharks, and great hammerheads.

AVERAGE WATER TEMP: 81°F (27°C) **AVERAGE VISIBILITY:** 75 feet (22.9 m)
AVERAGE DEPTH: 50 feet (15 m) **TYPE OF DIVE:** Shark

Tiger sharks are showstoppers. Great white sharks are charismatic, but tigers are beauties, with elongated bodies averaging 12 feet (3.7 m) in length, patterned with distinctive dark vertical stripes. They exude *presence*—a concealed power and allure. And the Bahamas' Tiger Beach is the only place in the world divers can share clear, warm waters with these extraordinary creatures, as well as great hammerheads, lemon sharks, sand tiger sharks, and more.

Tiger Beach is actually a shallow bank, 30 to 70 feet (9 to 21.3 m) deep, about 25 miles (40.2 km) north of Grand Bahama. (It's closer to West Palm Beach than Nassau.) The Gulf Stream that flows through the area also brings in mass congregations of large sharks, and divers eager to see them.

Baits are used to attract sharks in the area to the site, and once a tiger shark is spotted, it's game on. After the chain mail–clad feeder gets established on the ocean floor with a crate of fish (to keep the sharks' attention off descending divers), divers strap on more weight than usual to descend quickly through the clear waters to the white sand ocean floor, which is whipped up into even ridges by the present currents.

After getting fixed in position behind the feeder, the only thing left to do is watch the show. Along with the tiger sharks, oceanic whitetips, Caribbean reef, lemon, and sand tiger sharks are also usually present, a sharky melee of anywhere from five to 15 sharks gliding in deliberate patterns, swimming large loops with their eyes fixed on the feeding crate.

The operators who have been visiting Tiger Beach for years talk about "their" sharks in a way that might seem unusual to divers—at first. They speak with affection of individual personalities and peccadilloes, and that affection is contagious. Although the site

The beach's namesake shark patrols the reef.

Among other species, a lionfish shelters in the coral.

A tiger shark shows off its gigantic and powerful jaws.

That's why Tiger Beach is a challenging dive. Air consumption and decompression are rarely an issue. What's at hand is being comfortable in an environment with unpredictable and inherently dangerous animals. But that's exactly why divers come here: to see wild animals in a wild place, exactly how they should be—free. ○

"Interested in sharks? This is Mecca.
There are so many species in one fell swoop.
On almost any given day of the year,
you'll be surrounded by big tiger sharks
in 20 feet of water."

—BRIAN SKERRY

WASHINGTON

TITAN MISSILE SILO

Indulge your inner James Bond and dive a former Cold War missile silo.

AVERAGE WATER TEMP: 55°F (12.8°C) **AVERAGE VISIBILITY:** Good, unless silt is kicked up
AVERAGE DEPTH: 110 feet (33.5 m) **TYPE OF DIVE:** Missile silo

A 10-minute drive from Royal City, Washington, and 154 miles (247.8 km) southeast of Seattle, a Cold War secret lies underneath the Norman Rockwell farming community in the heart of the Columbia Basin.

The Titan missile silo is one of many weapons silos the government built throughout the United States. After it was decommissioned in the 1960s, it was flooded and became a recreational dive site.

Jaded on wrecks? Bored with cenotes? Over coral reefs? A missile silo might be just the new experience you're looking for. There's nothing to see in the way of wildlife, but you will get a firsthand glimpse of the Cold War.

Divers prepare by setting up their gear on wooden decks before entering the aptly named "ready room," the staging area for the dive, which can be reached by following a ladder down. (Most divers opt to wear their gear, rather than carrying it by hand, but it can also be lowered by rope.) Once kitted up and ready, the walk—or wade—to the silo begins—water fills the tunnel to knee or waist level.

The silo that once housed a Titan I intercontinental ballistic nuclear missile is 160 feet (48.8 m) deep, 100 feet (30 m) of which are flooded, creating a deep dive, and also a night dive, as there's no additional light. Although no weapons are left to be seen, a dive through the silo is pure novelty. Prepare to descend, an easy cruise straight into history.

The dive isn't difficult, but it requires prerequisites, including advanced and night diving certification, plus cave diving qualifications if you'd like to explore side tunnels. You'll also need the right gear (a minimum of two lights, for example). There are no weather considerations, but the water is always cold. ❍

What You'll See: A Cold War Relic

A dive in the Titan Missile Silo
is a swim through history

MADAGASCAR

NOSY BE

Topaz waters filled with sharks, whales, fish, and turtles

AVERAGE WATER TEMP: 83°F (28.3°C) **AVERAGE VISIBILITY:** 75 feet (22.9 m)
AVERAGE DEPTH: 50 feet (15 m) **TYPE OF DIVE:** Open water, reef, wall, and drift

Madagascar is often called the "land that time forgot." This Indian Ocean island located off the east coast of Mozambique is an impressive collection of biodiversity from stately baobab trees to quirky lemurs. Due to its long separation from the African continent, three-fourths of Madagascar's flora and fauna aren't found anywhere else on the planet.

Tucked off the northwest nose of this fantastical island is another one—Nosy Be, surrounded by topaz waters teeming with life: more than 56 types of sharks (including zebra, scalloped hammerheads, and tawny nurse), five of the seven species of marine turtle (including loggerhead and leatherback), 1,300 types of fish (including parrotfish, black-spotted sweetlips, and groupers), and more than 34 types of whales and dolphins (spinner, pantropical, and humpback, among others). Every year, around 10 percent of the world's humpback whale population migrates to Madagascar and can be viewed from Nosy Be.

The "big island"—the largest in the Mozambique Channel archipelago—is the perfect base for divers looking to explore and make the most of this unique location with its wide variety of dive sites: reef, wall, drift, and wreck.

Manta Point attracts its namesake species, particularly in September, when they aggregate to feed on plankton. Schools of unicornfish and humphead parrotfish wind their way around sea fans, while smaller creatures like garden eels, gobies, and shrimp nose around the sandy bottom and nooks and crannies.

Atlantis Point also attracts manta rays, as well as green and hawksbill turtles, eagle rays, and zebra sharks. The flat reef gives them different environments to explore.

Charlie's Point lies in the warm-water channel between Nosy Be and nearby Nosy Sakatia. Its sheltered location makes it ideal for smaller creatures (Spanish dancers, seahorses, lionfish, sponge crabs, and giant sea cucumbers), as well as diver training and night dives.

A leatherback sea turtle cruises the sapphire waters.

Schooling fish swarm a wreck in front of a diver.

What You'll See: Oceanic Whitetip, Scalloped Hammerhead, Tawny Nurse, and Zebra Sharks • Loggerhead, Olive Ridley, Leatherback, Green, and Hawksbill Turtles • Spinner and Pantropical Dolphins • Humpback Whales • Manta Rays • Blackspotted Sweetlips • Semicircle Angelfish • Sea Cucumbers • Spanish Dancers • Eagle Rays • Angelfish • Butterflyfish

Nosy Kisimasy is tucked off the southeast tip of Nosy Be, another sheltered spot with a tranquil reef, only 50 feet (15 m) deep and free from crowds—it's not a very well-known spot. Rich coral gardens attract hawksbill and green turtles, semicircle angelfish, clownfish, eagle rays, and butterflyfish.

A short boat ride is required to reach Nosy Tanikely, an island and marine park south of Nosy Be, between the island and Madagascar's northern coast. This is a popular spot, with some of the area's best marine life—bright anemones, sea stars, seahorses, and turtles in crystalline water. Delve deeper and you might see a guitarfish and schools of sweetlips and two-spot snapper.

Nearby, Seven Little Sharks is a drop-off with superb visibility, more than enough to see its namesakes: hammerheads, tawny nurses, and oceanic whitetips. ○

Travel Tip:

May through December is the best time to visit Madagascar, with October to December the best months for diving. Avoid cyclone season, which runs from January to March. Year-round, diving is usually better in the mornings, with the wind strengthening in the afternoons.

INDONESIA

MIKE'S POINT

A wall dive with large pelagics, coral, and macro species

AVERAGE WATER TEMP: 82°F (27.8°C) **AVERAGE VISIBILITY:** 150 feet (45.7 m)
AVERAGE DEPTH: 58 feet (17.7 m) **TYPE OF DIVE:** Wall

Bunaken is a small island in a chain of islands on the Celebes Sea. Located off the north coast of North Sulawesi, Indonesia, the Bunaken National Marine Park was one of Indonesia's first—established in 1991, and desperately needed.

Rich in biodiversity and a migratory highway for sperm whales and other marine life, the area was rapidly being fished to death. The establishment of the nearly 200,000-acre (809.4 sq km) park (97 percent of which is sea), as well as the divers who come to enjoy projected areas, is helping provide alternative employment for fishermen.

North Sulawesi's lack of continental shelf means the coastal area plummets directly to the seafloor, which makes for interesting diving. Mike's Point is one of the approximately 25 dive sites in Bunaken, a drift dive accessible by shore that sweeps divers in a strong current along a wall the shape of an amphitheater.

On the wall are sea fans and black coral, schooling baitfish, bumphead parrotfish, and the flash of the occasional silvertip shark.

This is another dive destination for adventurous travelers willing to explore off the grid, but the colorful, steep reefs flourishing with life are worth the effort.

Mike's Point is a year-round dive site with the best conditions from March to October. November through February can bring rougher surface conditions. All divers must pay an entrance fee (approximately $6 U.S. a day), which helps support the park. ❍

"Every other breath we take is from the ocean."

—BRIAN SKERRY

What You'll See: Sea Fans • Black Coral • Barracuda • Dugongs • Silvertip Sharks • Bumphead Parrotfish • Angelfish • Whitetip Reef Sharks • Tilefish • Eagle Rays

A stunning sunset falls behind equally beautiful coral reefs and sea stars.

SOUTH AFRICA

ALIWAL SHOAL

The most adrenaline-pumping dive in the world

AVERAGE WATER TEMP: 75°F (23.9°C) **AVERAGE VISIBILITY:** 30 feet (9 m)
AVERAGE DEPTH: 80 feet (24 m) **TYPE OF DIVE:** Wreck and sardine run

Tighten your weight belts. The sardine run is about to begin.

For divers, filmmakers, and photographers who have experienced the heart-stopping fracas of the annual Aliwal Shoal sardine run, no other dive comes close.

Once a year between May and July, millions and millions of sardines follow cold currents in close proximity to the shore. No one knows why they do this, but it's an impressive sight. The shoal of sardines can stretch two miles (3.2 km) wide and 10 miles (16 km) long, filling a water column 100 feet (30 m) deep as they travel hundreds of miles up the coastline.

This kicks off a frenzy of activity. The bait balls are unpredictable, identified only by the boiling of water as predators swifter to the mark than scuba divers discover its appearance.

Cape gannets dive-bomb the surface while underneath, the quicksilver fish are picked off by Cape fur seals, common dolphins, blue marlin, and sharks—lots and lots of sharks: hammerhead, ragged-tooth, tiger, bull, silky, blacktip, and even the occasional great white.

A humpback whale shows its underbelly as it jumps from the water.

What You'll See: Hammerheads • Ragged-Tooth, Tiger, Bull, Silky, Oceanic Blacktip, and Great White Sharks • Blue Marlin • Common and Bottlenose Dolphins • Bryde's Whales • Scorpionfish • Cape Fur Seals • Frogfish • Manta Rays • Nudibranchs • Whale Sharks • Wrecks • Sardines

If a diver is lucky enough to be on hand, the chaos—and strange symmetry—is mind-bending, and often over in a flash.

The sardine run is unpredictable—all you can do is prepare, plan, and hope. This dive is not without risk (predators usually cannot see what's ahead of them), so follow instruction, especially if it's your first sardine outing.

Although the sardine run doesn't take place *at* Aliwal Shoal, this one-mile-long (1.6 km) stretch of reef has the cocktail of current the sardines seem to be looking for, which means they often pass by. The local dive operators have been diving this phenomenon for years, soaking up local knowledge that provides visiting divers the best opportunity to experience it for themselves.

Located offshore of Umkomaas in the Indian Ocean, about 30 miles (48.3 km) south of Durban, Aliwal Shoal is a lively dive destination in its own right. It lies within the north-south current in a marine protected area, sitting approximately 64 feet (19.5 m) deep and 2.5 miles (4 km) offshore, the remnants of an ancient sand dune colonized by coral species.

It's an unevenly built structure, with crevices, overhangs, nooks, and crannies, attracting more than 1,200 marine species. From whale sharks to nudibranchs, you'll find them at Aliwal Shoal.

Wrecks are also present. The British steamship S.S. *Nebo* sank in 1884, traveling from England to Durban, resting upside down and partially covered in 89 feet (27 m) of water. Not much remains after the inexorable ebb and flow of tides except for her propellers and boiler, and plentiful marine life, including barracuda, Queensland groupers, nudibranchs, guitarfish, and crocodilefish.

The Norwegian carrier *The Produce* is another local wreck. It sank in 1974 and sits slightly deeper in 100 feet (30 m) of water, more intact than the S.S. *Nebo,* with resident moray eels, lionfish, and the rare harlequin goldies.

Still craving adrenaline? Aliwal Shoal has one more trick up its sleeve: a 60-minute baited shark dive attracting anywhere from six to 50 oceanic blacktip, dusky, bull, and tiger sharks in 26 feet (7.9 m) of water. ❍

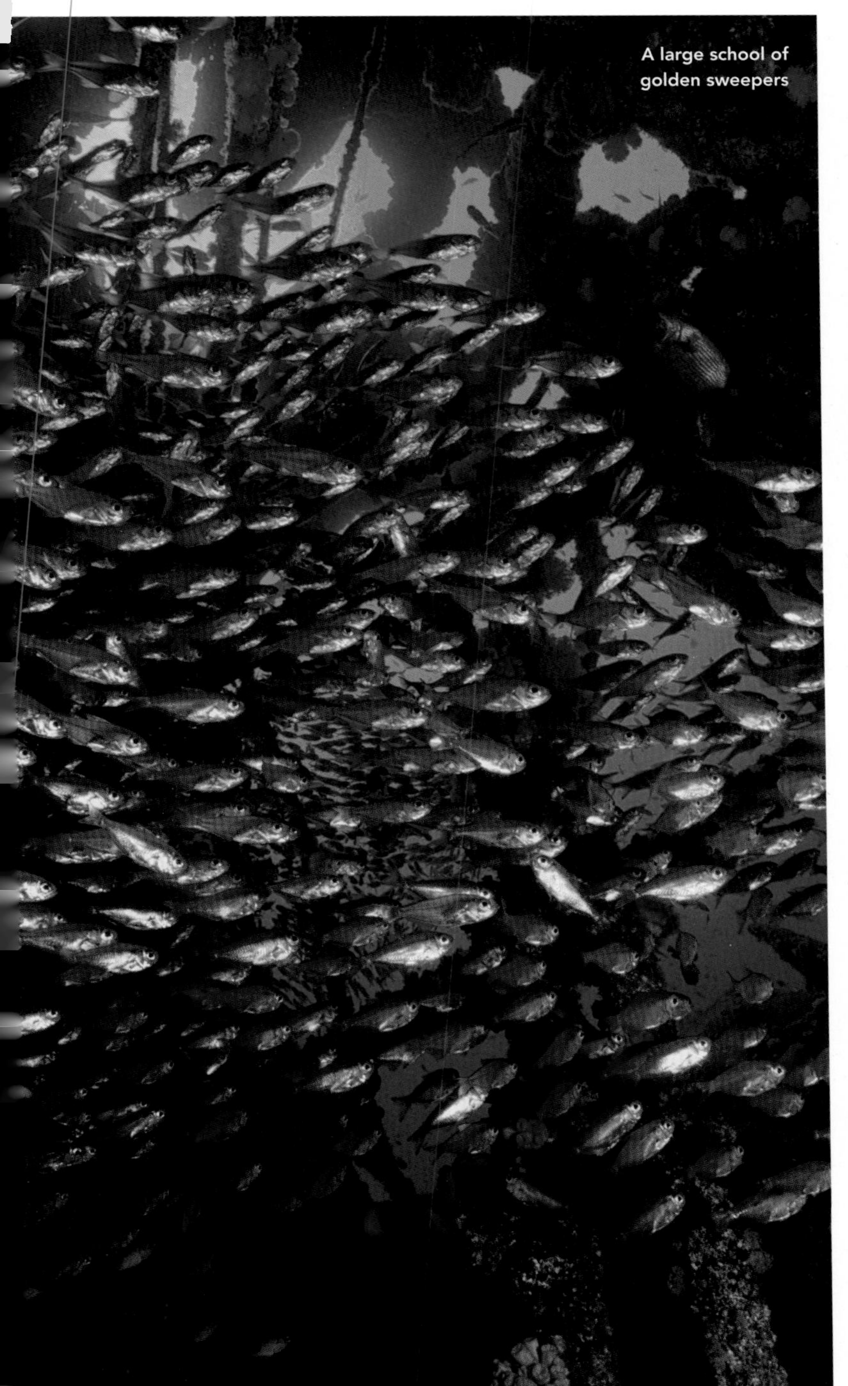
A large school of golden sweepers

A green sea turtle stops for a bite.

A raggy scorpionfish lies motionless on the coral reef.

A swordfish hunts a shoal of sardines.

SCOTLAND

SCAPA FLOW

Dive through history via seven German WWI shipwrecks.

AVERAGE WATER TEMP: 48°F (8.9°C) **AVERAGE VISIBILITY:** 65 feet (19.8 m)
AVERAGE DEPTH: 112 feet (34 m) **TYPE OF DIVE:** Wreck

Scapa Flow has watched the golden age of shipping sail in and out of its 120-square-mile (310.8 sq km) harbor since the Vikings settled here in the late eighth century. Even to untrained eyes, a map reveals what a strategic harbor Scapa Flow was. The Orkney Islands, an archipelago resting just out of reach of the northern tip of the Scottish mainland, surround a vast open ocean area, accessible only via a maze of narrow channels. It was an ideal place to shelter—or hide—a fleet.

In 1918, Scapa Flow was selected as a holding pen for the German fleet following the armistice of WWI. Admiral Ludwig von Reuter duly placed 74 ships—50 destroyers, 10 battleships, eight light cruisers, and six battle cruisers—into Scapa Flow to await terms, but he had no intention of handing over his ships as spoils of war.

On June 21, 1919, he ordered the ships be scuttled, sinking the entire fleet, an act that claimed some of the last casualties of WWI.

The British Royal Navy scrambled, preventing 22 ships from sinking, but 52 slipped from their grasp. Forty-five of those were salvaged and raised from

A mostly intact WWI shipwreck lies on the seafloor.

What You'll See: Wrecks • Conger Eels • Sea Urchins • Sea Stars • Brittle Stars • Sponges • Crabs • Jellyfish

their watery graves (an action wreck divers still lament) during the time between WWI and WWII. Three battleships and four light cruisers were left behind.

During WWII, Scapa's passages were blockaded to protect the British fleet, but a German submarine still managed to sneak through, claiming the H.M.S. *Royal Oak* and adding another ship to Scapa Flow's depths.

So Scapa Flow's wreck diving reputation is understandable. The seven remnants of the German fleet rest near the island of Cava: three battleships (*Kronprinz Wilhelm, König,* and *Markgraf*), each stretching more than 550 feet (167.6 m) in length, a staggering size that never fails to impress divers. They lie upside down in 80 to 150 feet (24 to 45.7 m) of water, presenting their upturned hulls (and massive 12-inch/30.5 cm-diameter guns) to visiting divers, an awesome and unforgettable armament.

The smaller cruisers *(Coln, Karlsruhe,* and *Dresden)* measure around 510 feet (155.4 m) in length, resting in 45 to 120 feet (13.7 to 36.6 m) of water, near the mine-layer *Brummer.* These ships fell on their sides and are more accessible than the battleships. Dangling anchor chains are a sober reminder of the origin of these wrecks. The *Brummer,* in particular, is intact, with its imposing weaponry and a brass bridge that has to be one of the highlights of Scapa Flow.

Crabs and lobster peak out of rents in the ships' hulls, while schooling pollock mill about the decks, but marine life is muted here. The wrecks demand the attention, a solemn marker of maritime history, those who made it home, and those who didn't.

Don't forget to tear your eyes away from the wrecks long enough to admire to Scapa Flow's stunning and oh-so-Scottish scenery, with heather-covered hills, rich farmland, and dramatic coastlines. Scapa Flow is home to unique treasures both above and below the sea, from the cheerful purple and gold Scottish primrose, found only in northern Scotland, to the giant and rare flapper skate, fished to near extinction, and the even bigger basking shark, which visit the area during the summer months.

Summer (June through September) is the nicer time to dive Scapa Flow. Make sure to visit the Stromness Museum during your stay. It has a wide variety of artifacts and exhibitions to enhance the Scapa Flow experience. ❍

Dock shrimp feed on herring roe.

A white-spotted sea anemone

A decorated warbonnet fish stares at its visitor.

A topside wreck of an old blockship

K980

FLORIDA

DEVIL'S DEN

An underground prehistoric spring, rich in fossils

AVERAGE WATER TEMP: 72°F (22.2°C) **AVERAGE VISIBILITY:** 95 feet (29 m)
AVERAGE DEPTH: 54 feet (16.5 m) **TYPE OF DIVE:** Freshwater spring

On cold Florida mornings, the steam can be seen rising from below the ground. This is the entrance to the Devil's Den, and the origin of its name.

The Sunshine State's only underground, prehistoric spring is located south of Gainesville and north of Ocala, a natural volcanic cave flooded by spring water at a steady and temperate 72°F (22.2°C), pleasant enough for a 3mm wet suit.

Divers and snorkelers are welcome to explore Devil's Den, a 120-foot-wide (36.6 m) cavern 54 feet (16.5 m) deep that is tucked away on a private rural property.

The water is crystalline, clear enough to see fossils embedded in the limestone walls. (Midday is best for light.) These remnants reach back to the Pleistocene age, 33 million years ago: Ground sloth, saber-toothed cats, and mastodon fossils have been found in Devil's Den. Although divers aren't allowed to remove any rocks or fossils from this area (unlike Cooper River, page 268), they can visit the University of Florida's Museum of Natural History in Gainesville, a 25-minute drive from Devil's Den, to view fossils collected in the spring. ○

"Exploring every limestone swim-through and shelf reveals fossil imprints on the walls and silty grim reaper signs, warning divers to stay out of dangerous caves. Although it is an open-water dive, rocks overhead give open-water divers a taste of cave diving while still having the sunlit surface safely nearby."

—JENNIFER ADLER OWEN, CONSERVATION PHOTOGRAPHER, CAVE DIVER, AND EDUCATOR

What You'll See: A Turtle Named Nelson • Catfish • Rock Formations • Fossils

Divers prepare to descend from the landing dock of Devil's Den.

COSTA RICA

ISLAS SANTA CATALINA

Big fish action near underwater volcanic formations

AVERAGE WATER TEMP: 82°F (27.8°C) **AVERAGE VISIBILITY:** 50 feet (15 m)
AVERAGE DEPTH: 75 feet (22.9 m) **TYPE OF DIVE:** Open water

Islas Santa Catalina is an archipelago of uninhabited bush-clad rock islets rising out of the Pacific Ocean less than 15 miles (24 km) from the northwest coast of Costa Rica. Their location on the edge of the open ocean, paired with nutrient-rich currents, makes Islas Santa Catalina a magnet for large pelagic species including bull sharks, as well as multitudes of rays, and the giant manta ray, the world's largest ray with wingspans stretching 23 feet (7 m).

This is a changeable area, which is why it's not suited for beginners. Visibility can vary, and conditions deteriorate without warning. Delicate coral reefs aren't the attraction here: Divers come to see the big marine life, as well as the striking volcanic rock formations—caverns, arches, and tunnels, as well as the few scattered shipwrecks.

Two dive locations get top billing: The Wall (40 to 70 feet/12.2 to 21 m) is a sharky channel also frequented by mantas with strong surge and occasional currents. Divers descend through clouds of tropical fish (pufferfish, butterflyfish, angelfish) and swim through a channel that is home to whitetip reef sharks and octopuses before drifting past a series of pinnacles. Rays of all kinds—spotted eagle, cow-nose, mobular—swoop past, and blocking out the overhead sunlight are the unmistakable wings of a manta ray, searching for a cleaning station to visit.

The Point (40 to 110 feet/12.2 to 33.5 m) is home to two large cleaning stations. Divers descend in the shallows of an island and allow the current to sweep them out to deeper waters. This is another good location to see both manta rays and whitetip reef sharks, as is Big Catalina.

A pufferfish rests in the rocky sand.

Two remora hitch a ride
on a giant manta.

What You'll See: Giant Manta Rays • Bull, Whitetip Reef, and Tiger Sharks • Eagle, Devil, Cow-nose, Spotted Eagle, Bullseye, and Mobula Rays • Frogfish • Butterflyfish • Pufferfish • Angelfish • Whale Sharks • Humpback Whales • Orcas

Some of the other popular dive locations include Two Hats (renowned for its wall diving), Dirty Rock (ironically a great spot for visibility), Big Cupcake (angelfish are plentiful here), and Little Cupcake (multitudes of brightly colored parrotfish).

Islas Santa Catalina, also known as the Catalina Islands, or "the Cats," is one of those unique places where divers can experience an iconic moment they've seen captured in underwater photographs. Here you'll find multitudes of rays in swarms of 15, 20, or 30 silhouetted against the sun like a winged unit of floating, flying handkerchiefs, graceful and ethereal.

Easily accessible by the resort towns in the northern Guanacaste Province, don't underestimate a day trip to Islas Santa Catalina. This is a remote area with quickly changing conditions, so pack rain gear, several bottles of water, sunscreen, insect repellent, and everything else you might need in a just-in-case situation, along with your dive gear. ❍

Travel Tip:

May to November is the rainy season, with daily showers (only lasting for a few hours) and increased winds. October is the month to avoid. December through April is the dry season, with the pelagics replaced by schools of fish, macro, and manta rays.

EGYPT

S.S. *THISTLEGORM*

One of the world's most popular wrecks—with good reason

AVERAGE WATER TEMP: 77°F (25°C) **AVERAGE VISIBILITY:** 66 feet (20 m)
AVERAGE DEPTH: 79 feet (24 m) **TYPE OF DIVE:** Wreck

The S.S. *Thistlegorm,* a 415-foot (126.5 m) WWII steamship, is a wreck-diving trifecta, hence its popularity: It is within recreational limits, it's well preserved, and it is a time capsule of artifacts.

Located on the bottom of the Red Sea, the S.S. *Thistlegorm* lies broken in two pieces, the result of an aircraft bombing in October 1941. The British-built steamship had been drafted into the war effort, fitted with antiaircraft and machine guns, and dispatched to deliver desperately needed equipment (motorcycles, motor parts, locomotives) to northern Africa. When it was struck, the bombs ripped the ship in half.

The stern lies on its port side around 105 feet (32 m), with its enormous propeller and antiaircraft guns mounted on the deck, while the bow is shallower (52 feet/15.8 m), sitting upright. The ship is largely intact (barring the impact point), making it an underwater museum, which is what attracts divers. Mark II Bren carrier tanks, Bedford trucks, Enfield carbines, Bren guns, Norton motorcycles, even pairs of Wellington boots are on underwater display, a watery reminder of the moving parts of war.

Tear your eyes away from the wreck and you're likely to see a large variety of marine life, too. Nearby swimmers may include trevallies, batfish, and hawksbill turtles.

On a visit to the wreck, expect crowds and strong currents, and be mindful of deep diving in a remote area. A liveaboard allows the option of diving the site in the morning before the day boats arrive from Sharm el Sheikh. Before diving the S.S. *Thistlegorm,* make sure you're comfortable with diving wrecks—it has open holds and dark, enclosed interiors, so experience in this environment will add to the enjoyment. ❍

What You'll See: Hawksbill Turtles • Soldierfish • Trevallies • Batfish • Lionfish • Blackspotted Sweetlips • Fusiliers • Crocodilefish

Fish swarm a motorcycle, more than 75 years after its carrier ship sank in WWII.

EGYPT

ELPHINSTONE REEF

A wall dive in the Red Sea with sharks

AVERAGE WATER TEMP: 80°F (26.7°C) **AVERAGE VISIBILITY:** 87 feet (26.5 m)
AVERAGE DEPTH: 73 feet (22.3 m) **TYPE OF DIVE:** Wall

Elphinstone Reef lies in the open ocean of the Red Sea, just about 7.5 miles (12 km) off Egypt's Marsa Alam coast. A few out-of-place whitecapped waves are the reef's only tell—the only indication that something lies beneath.

Elphinstone's defining characteristic is its torpedo shape: Running about 1,000 feet (305 m) in length and only 100 feet (30 m) wide, tapering off at both ends, it's bordered by two underwater plateaus at the northern and southern tips, with plunging straight walls on the western and eastern sides. Its unique shape and open ocean location offer some spectacular drift dives along steep walls, making it one of the premier dive destinations in the Red Sea.

What You'll See: Giant Trevallies • Fusiliers • Hammerheads • Tiger, Gray Reef, Silky, Silvertip, Thresher, and Oceanic Whitetip Sharks • Dogtooth Tuna • Manta Rays • Napoleon Wrasses • Gorgonians • Sea Whips • Sponges

Multihued coral and colorful fish abound in the Elphinstone Reef.

A small cluster of lemon butterflyfish.

The Northern Plateau starts around 66 feet (20 m), gently sloping to a drop-off. This 300-foot-wide (91.4 m) plateau is coated in sea whips, sponges, gorgonians, and soft corals in purples and oranges, with angelfish, triggerfish, giant trevallies, and barracuda swimming alongside. This is the most likely place to spot hammerheads, so keep an eye out toward the open ocean. These beauties aren't common, but they are around from time to time, depending on the season, and many divers are so transfixed by the reef's bursts of color and clouds of fish they forget to look over their shoulders.

The Southern Plateau is similarly laid out and is a favorite for curious oceanic whitetip sharks engulfed in an entourage of pilot fish. If you see one, count yourself lucky—not

"Sharks are sculpted by nature to be perfect, move exquisitely through the water, graceful and powerful, without wasting a bit of energy."

—BRIAN SKERRY

many divers get the opportunity. At the time of writing, the oceanic whitetip is listed as "threatened" under the Endangered Species Act—the first shark species to receive this protection in Atlantic waters. It is also listed as vulnerable by the World Conservation Union. At one time, not too long ago, these sleekly silhouetted creatures were legions strong, in numbers so great no one even bothered to count them. Now, decimated by the shark fin trade and commercial fishing (research suggests populations have declined

Showing off its enormous size, a Napoleon wrasse swims by a diver.

Travel Tip:

Elphinstone Reef can be dived all year round. May to August is manta ray season, October through December is best for oceanic whitetips, with spring and fall the time to spot other sharks. The reef is accessible by boat, either a day trip from Marsa Alam (roughly a 40-minute ride) or a liveaboard. It's a popular spot, so you'll be sharing it with other divers.

by 70 to 80 percent), these bold characters are a rare and wonderful thing to glimpse. And if you're lucky, you may get to do so at Elphinstone.

Another unique sighting on the Southern Plateau is the Elphinstone Arch, a 33-foot-arch (10 m) connecting the east wall with the west. It's tech-diver territory, too deep for recreational limits, and the sweeping current has surprised more than one diver. Legend has it that an Egyptian pharaoh was buried under this sarcophagus arch, which is largely encrusted with coral.

On the sides of the torpedo, the Western Wall has caverns and crevices adding texture to its steep wall. Snapper, triggerfish, and soldierfish seem to like this spot.

The Eastern Wall is a drift dive worth dreaming about, with strong currents carrying divers past colorful coral and a teeming population of Napoleon wrasses, barracuda, jacks, and angelfish.

Elphinstone is one of the sharkiest spots in the Red Sea. Several species have been sighted here, including gray reef, blacktip reef, oceanic whitetip, hammerhead, and even thresher sharks. Spring and autumn are the best times of year to see them, so keep your eyes peeled and facing away from the reef into the open ocean. Late summer and early autumn bring the warmest water temperature—but avoid August, when dive crowds are at their heaviest.

The unpredictable and strong currents are what make Elphinstone Reef a more challenging dive. Many tour operators require divers to have a certain number of dives prior to diving Elphinstone (pause for thought if they don't), and divers should ensure they're not only comfortable in current, but that they factor into their dive plan what to do if they get carried off the reef. It's unlikely, but it never hurts to be prepared. ❍

Pilot fish accompany an oceanic whitetip shark on its swim.

NEWFOUNDLAND

BELL ISLAND

A historical tour of wrecks, abandoned mine shafts, and a desperate moment in time

AVERAGE WATER TEMP: 45°F (7.2°C) **AVERAGE VISIBILITY:** 65 feet (19.8 m)
AVERAGE DEPTH: Up to 150 feet (45.7 m) **TYPE OF DIVE:** Wreck and mine shaft

Newfoundland's Bell Island is a bean in Conception Bay, wrapped up in the arms of the Avalon Peninsula in the middle of the North Atlantic Ocean. Only 11 square miles (28.5 sq km) in size, it's difficult to understand how this tiny speck could attract the Germans' attention during WWII, but it did—twice. U-boats were sent in to cripple iron ore production, the economic powerhouse of the region.

In two attacks, the Germans sank the S.S. *Rose Castle,* S.S. *Saganaga,* S.S. *Lord Strathcona,* and the Free French vessel S.S. *PLM 27.*

Then, in 1966, after a postwar steady market decline, the 1,800-foot-deep (548.6 m) mines that made up the island's economy were flooded—without the workers being notified in advance. In one fell swoop, the majority of Bell Island's 13,000 residents found themselves unemployed. More than 50 years later, Bell Island (population 2,500) is still searching for an industry and identity.

That turning point remains frozen in time, and underwater. Divers can explore the wrecks, although the cold water, isolation, and depths require extra mindfulness. Far from being rusted-out shells, they're covered in vibrant cold-water colors: whites, oranges, golds, and pinks, a collection of anemones, sea stars, and lion's mane jellyfish.

The mine itself is a time capsule of economic collapse. Divers who explore these parts will find left-behind leather shoes and broken shovels due to be repaired, among other artifacts. This is a full technical dive, requiring proper training and certification.

With its soaring cliffs overlooking the wild ocean, Bell Island is a breathtaking place to visit. On land, stop at the No. 2 Mine and Museum to learn more about its history. ❍

What You'll See: Wrecks • Flooded Mine Shaft • Anemones • Sea Stars • Jellyfish

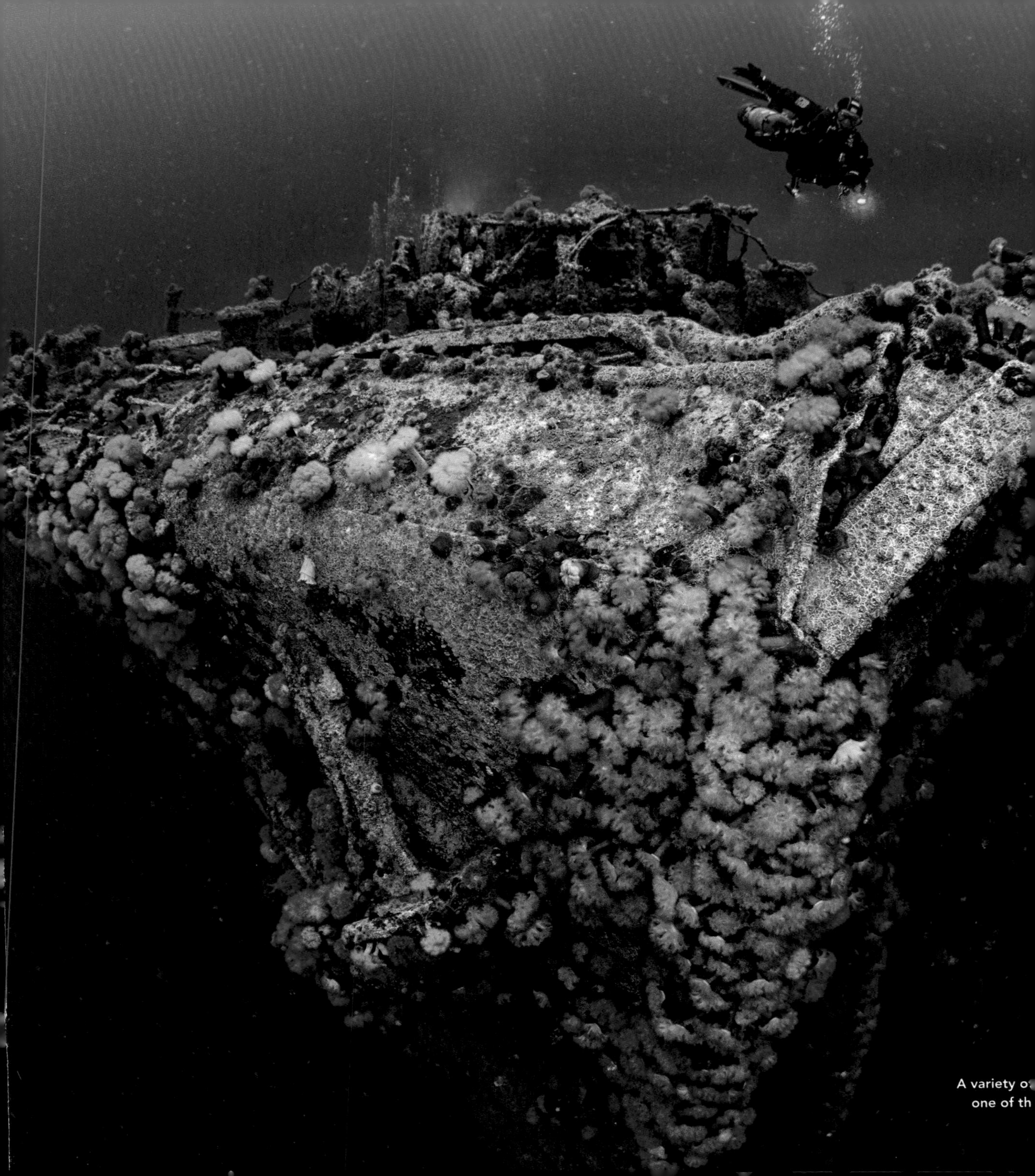

A variety o
one of th

PART THREE

ADVANCED AND ALL-LEVEL DIVES

Braving the cold, a diver swims beneath Antarctica's ice.

AUSTRALIA

NINGALOO REEF

The world's largest fringe reef is a pelagic paradise.

AVERAGE WATER TEMP: 70°F (21°C) **AVERAGE VISIBILITY:** 16 to 80 feet (4.9 to 24 m)
AVERAGE DEPTH: 50 to 90 feet (15.2 to 27.4 m) **TYPE OF DIVE:** All level—reef

It's mind-boggling to think that one of the driest landscapes in Australia is also one of its best dive sites. Ningaloo, a 1.7-million-acre (6,880 sq km) UNESCO World Heritage site, receives a scant nine inches (23 cm) of annual rainfall, its arid red desert giving way to white sand beaches and the emerald of the Indian Ocean, color so vivid it stings the eyes. For 30,000 years, this spur of land on Western Australia's northwest coast has attracted visitors, from Aboriginal peoples to Dutch captain and explorer Willem Jansz (1618), and now divers.

Divers love Ningaloo for two reasons. Its 160-mile (257.5 km) fringe reef is the world's largest, positioned one to two and a half miles (1.6 to 4 km) offshore. Its peninsula shape and islands offer a wide variety of dive sites, from shallow (52 feet/15.8 m) shore dives, to outer reef dives (90 feet/27.4 m) with schooling pelagics, to swim-through channels near the Muiron Islands, an hour north by boat.

And then there's Ningaloo's abundant marine life. Western Australia is a migration highway, with whale sharks, humpback whales, and manta rays all putting in seasonal appearances, not to mention regular sightings of sharks (blacktip reef, whitetip reef, tawny nurse, wobbegong, and zebra), turtles, and even dugongs. With more than 300 species of soft coral (producing mass spawning events), 600 species of mollusks, and 500 species of fish, including angelfish, batfish, lionfish, and schools of snapper, divers experience the thrill of big pelagics as well as teeming reef life.

The only issue with Ningaloo can be varied visibility, ranging from 16 to 80 feet (4.9 to 24 m) depending on winds, tides, and weather patterns. However, given the large number of dive sites around the peninsula, most operators find it easy to work around any visibility issues.

An aerial view of
Ningaloo Reef

Ningaloo is famous for seasonal appearances of these gentle giants.

What You'll See: Whale Sharks • Humpback Whales • Blacktip Reef Sharks • Whitetip Sharks • Zebra Sharks • Tawny Nurse Sharks • Manta Rays • Boulder, Cabbage, Brain, Branching, and Staghorn Coral

Popular dives include Navy Pier (see page 154), Bundegi Sanctuary (a shallow dive in a nursery area for a variety of species), and the Muiron Islands (an hour by boat with a maximum depth of 66 feet/20 m and suitable for all divers, with swim-throughs, large pelagics, and coral gardens).

And if that isn't enough, don a snorkel and fins and cruise the warm, clear waters with some of the most impressive creatures on the planet: whale sharks and humpback whales.

The opportunity to snorkel with these gentle giants is rare, and Ningaloo Reef is one of the few places on Earth where they appear regularly. The currents, nutrient-rich water, and warm temperatures make this an ideal spot for these fascinating creatures. Whale sharks, reaching 35 feet (10.7 m) in length, migrate here between March and August. Looking for something slightly bigger? Humpback whales (which can reach up to 60 feet/18 m long) congregate between August and November. ❍

Travel Tip:

Ningaloo is a remote area, located 750 miles (1,207 km) north of Perth. Its population surges during the summer months (December through February), with sun-seeking travelers reveling in calm seas and temperatures ranging from the mid-80s to over 100°F (37.8°C). Book well in advance during peak seasonal times, particularly whale shark season.

FRANCE

ÉMERGENCE DU RESSEL

A cave system for novice and experienced divers

AVERAGE WATER TEMP: 55°F (12.8°C) **AVERAGE VISIBILITY:** 330 feet (100.6 m)
AVERAGE DEPTH (FEET): 30 to 60 feet (9 to 18 m) **TYPE OF DIVE:** Advanced–cave

Just beyond the entrance of the Émergence du Ressel, the cave system forks: One passage is shallow (30 feet/9 m), while the other drops deeper (60 feet/18 m), eventually converging in a 2.5-mile (4 km) loop. Although most divers won't complete the entire loop, the passages are wide enough for back-mounted rigs, and the visibility (excluding silt kickup and other environmental influences) is a clear 330 feet (100.6 m). It's a good cave for novice cavern divers to explore close to the entrance, but for divers with more technical experience and preparation, there's so much more, attracting cave diving experts from around the world.

Large slabs of sand-colored rock, rounded and carved into otherworldly shapes, surround divers, seemingly funneling daring explorers into the bowels of the Earth.

Located approximately halfway between Dordogne and Toulouse in southern France, Émergence du Ressel is part of a region known for impressive cliffs and limestone caves. This cave system starts in the bed of the River Céle. It is legendary for its exploration history. In 1999, two British divers completed a five-hour inward dive, followed by a six-hour outward dive. Over the next three years, they discovered five submerged cave passages.

April through October is the best time to visit. Although summer is a busier season for traveling in France, low rainfall leads to lower water levels and better visibility. The water in the cave system hovers around 55°F (12.8°C) year-round. Cavern diver certification is needed. This is a self-organized dive, so use caution. ❍

What You'll See: Massive Cave System ● Limestone Channels ● Dry Caves at Depth

Divers light their way through the complex cave system.

THAILAND

SIMILAN ISLANDS

The Andaman Sea's wonderland of hard and soft corals

AVERAGE WATER TEMP: 80°F (26.7°C) **AVERAGE VISIBILITY:** 100 feet (30 m)
AVERAGE DEPTH: 16 to 130 feet (4.9 to 39.6 m) **TYPE OF DIVE:** All level—liveaboard

Nine granite islands are strung along a 15-mile (24 km) chain in the azure blue of the Andaman Sea, off the western coast of Thailand. These are the Similan Islands, small slips of paradise with white coral sand beaches, jungle-clad hills, and oddly shaped boulders. It's an area so beautiful it's tempting to stay on land, exploring the oft-deserted beaches. However, you'll be missing out on a destination that regularly makes the top 10 list of the world's best dive sites.

The Similan Islands have more than 20 underwater Edens to explore. On the east coast, gently sloping coral reefs (mostly hard, some soft) host an abundance of marine life, such as ghost pipefish, ribbon eels, frogfish, clownfish, and a variety of nudibranchs. The diving here is relaxed and ideal for beginners, with gentle currents and 100-foot (30 m) visibility.

The western side of the islands has more teeth, with rocky gorges, swim-throughs, walls, caves, and

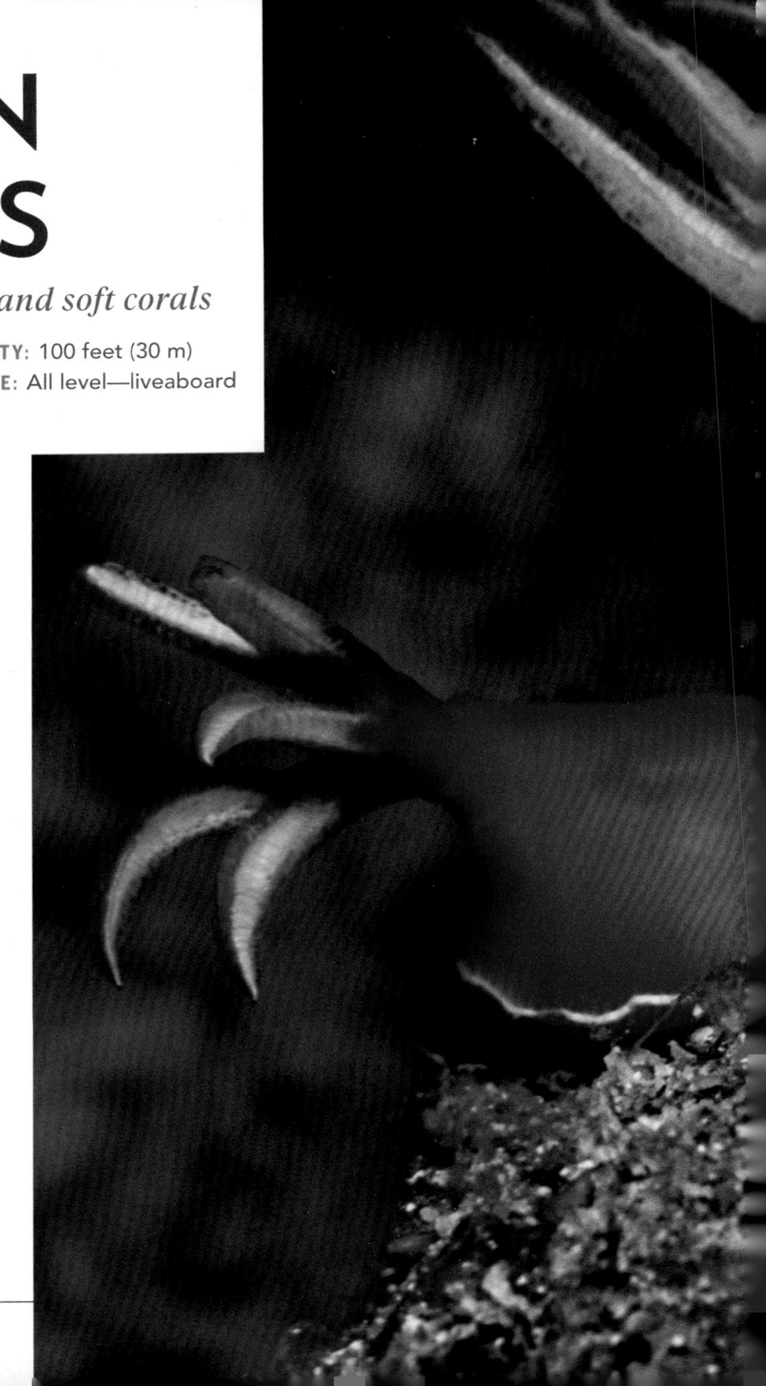

A Technicolor nudibranch rests in the coral.

What You'll See: More Than 500 Species of Hard and Soft Corals • Green, Hawksbill, Olive Ridley, and Leatherback Turtles • Whitetip and Blacktip Reef Sharks • Zebra Sharks • Whale Sharks • Nudibranchs • Triggerfish • Unicornfish • Clownfish • Guitarfish • Batfish

stronger currents. The dives here are deeper (dropping to 130 feet/39.6 m), and you're likely to see larger marine life, such as guitarfish, whale sharks, zebra sharks, manta rays, turtles (green, hawksbill, olive Ridley, and leatherbacks), batfish, moray eels, and giant fan corals.

Everywhere you look, there's something to see, from well-camouflaged octopuses, stonefish, cuttlefish, and scorpionfish, to delicate (but venomous) lionfish and petite seahorses.

The islands are located in Mu Ko Similan National Park, established in 1982. The name Similan means "nine" in the local language, referring to the number of islands within the park. Each island has a name but is usually referred to by a number. Number 8, for example, is Koh Similan, home to Donald Duck Bay, one of the most photographed collections of rocks in the world because of their resemblance to the cartoon character. This shallow dive spot is a favorite with snorkelers, as well as night divers.

Mu Ko Similan National Park closes for six months of the year (usually between mid-May and mid-October, but it can vary, so check before you go). This is due to the frequent tropical storms that sweep through during this time, but it also gives the area a much needed break from visitors. Several of the islands are permanently off-limits to protect egg-laying turtles.

The water is consistently warm throughout the year (around 80°F/26.7°C), but each season brings something new to the Similans. The strong currents at the end of February make for enjoyable drift dives and bring in bigger pelagics, like whale sharks and manta rays, while green and hawksbill turtles use the Similan Islands as a breeding ground from November to February.

At any time of year, this diverse, clear-water gem will prove a favorite, with something to interest divers of all levels. February to April presents the most stable weather conditions, with calm seas (now that the monsoon winds have ceased), clear skies, and visibility between 80 and 130 feet (24 to 39.6 m).

Liveaboards (four days/four nights) are the most popular way of visiting the Similan Islands, although longer excursions or day trips from Phuket are available. ❍

The Donald Duck Rock overlooks the crystal blue waters.

Porcelain crabs hide in the lush reefs of the Similans.

A striking red scorpion fish

A diver hovers, transfixed by the colors of the reef walls.

MEXICO

YUCATÁN CENOTES

Dive like Indiana Jones, exploring hidden wonders and remnants of an ancient civilization.

AVERAGE WATER TEMP: 77°F (25°C) **AVERAGE VISIBILITY:** 300-plus feet (91.4+ m)
AVERAGE DEPTH: Unknown **TYPE OF DIVE:** All level—cavern

First revered by the Maya people, cenotes are now worshipped by divers fascinated with the crystal clear waters (visibility 300-plus feet/91.4+ m), striking rock formations, and archaeological finds in these ancient sinkholes.

Cenotes stud Mexico's Yucatán Peninsula, a spur of land that kicks up into the Gulf of Mexico. Most can be found on the northwest coast, between Playa del Carmen and Tulum, the result of three of the world's longest underground water systems, which created the largest cave system in the world. Where cave ceilings have collapsed, cenotes appear, glistening emerald drops tucked in lush jungle, subterranean circles pierced only by dangling tree roots. Their eerily romantic, Indiana Jones-esque appearance makes them a draw for adventurers and explorers.

Anyone holding a basic scuba certification can dive cenotes. As long as there is natural light, cenote diving is considered cavern diving, rather than cave diving (which is also readily available here), and you can choose your level of difficulty, from novice to full-blown technical. New or claustrophobic divers should exercise caution and do their homework before taking the plunge, and the popularity of the area means there are plenty of quality dive shops to choose from.

Cenotes can be challenging to reach, often involving hauling dive gear along a jungle path, but the payoff is worth it. One giant stride drops you into another world: flooded, cavernous halls; stalagmites and stalactites; haloclines (the meeting of saltwater and freshwater); and misty rivers of hydrogen sulfide that can be up to 10 feet (3 m) thick. (Many of the cenotes have fixed guide ropes that divers can follow to help with navigation.)

Keep your eyes peeled and you might swim across an archaeological discovery. Ancient fossilized remains of coral, giant jaguars, mammoths, and 10,000-year-old human

Freshwater fish swim
between water lily leaves.

Among the many wonders of the cenotes are stunning stalagmites and cave structures.

Divers explore an underwater cave near Tulum.

Snails rest on a leaf in the still waters of a cenote.

What You'll See: Limestone Cave Systems ● Rock Formations ● Saltwater and Freshwater Fish (on occasion)

skeletons are just a few of the treasures found by cave divers, not to mention the occasional remnants of unlucky, wandering cows.

Popular cenotes include Dos Ojos (Two Eyes), two shallow, connected cenotes that are part of a larger cave section and suitable for divers of all levels, and the aptly named Pit, a 391-foot-deep (119.2 m) cenote with exceptional visibility (and stalactites) up to 100 feet (30 m), after which divers descend through a hydrogen sulfide cloud to encompassing darkness (130 feet/39.6 m).

With more than 6,000 cenotes on the Yucatán Peninsula, there are plenty to choose from, ranging from the frequently visited to the unexplored, a cenote to match the experience and adventurous spirit of every diver. ❍

BRITISH OVERSEAS TERRITORY

SOUTH GEORGIA

A siren call to adventure divers—wild, remote, unexplored

AVERAGE WATER TEMP: 35°F (1.7°C) **AVERAGE VISIBILITY:** 5 to 30 feet (1.5 to 9 m)
AVERAGE DEPTH: Less than 82 feet (25 m) **TYPE OF DIVE:** Advanced—cold water

Traveling to South Georgia requires patience, humor, and a healthy sense of adventure; diving South Georgia requires an even bigger dose of each of these qualities!

This barren island, three-fourths of which is perpetually covered in snow, is one of the world's harshest environments, supporting a few stubborn tundra plants, reindeer, penguins, and seals. The island is best known for British explorer Sir Ernest Shackleton's desperate 1916 crossing to rescue his crew. (This brief synopsis does not do this epic adventure justice. Read Shackleton's account in his book *South*.)

Barren Island, a mountainous 1,450 square miles (3,755.5 sq km) located 800 miles (1,287.5 km) from the Falkland Islands, is a challenging and virgin dive location. Although a few operators offer boat diving, it's cold (water temperatures are approximately 35°F/1.7°C) and surge-y, with five- to 30-foot (1.5 to 9 m) visibility. The diving is restricted to 82 feet (25 m) because help is a long way away, and dry suit experience and self-reliance are necessities.

You will be one of the first people to explore this 386,000-square-mile (999,735 sq km) marine protected area, lush with endemic macro species, sea stars, sea feathers, and nudibranchs. Antarctic fur seals are plentiful (and sometimes overly curious), and the island is pockmarked with sea caves.

Largely unexplored, this is a wild and naturally beautiful area for the adventurous who appreciate what it has to offer. On land, visiting the South Georgia Museum, located in Grytviken, in the former home of a whaling-station manager, is a must. It features exhibits on the island's history: Shackleton memorabilia, whaling, maritime, and natural history. ○

What You'll See: Antarctic Fur Seals • Unusual Macro Life • Sea Stars • Sea Feathers • Nudibranchs

An elephant seal comes ashore.

COLOMBIA

MALPELO

Sharks, sharks, and more sharks

AVERAGE WATER TEMP: 70°F (21°C) **AVERAGE VISIBILITY:** 30 to 100 feet (9 to 30 m) **AVERAGE DEPTH:** 100 feet (30 m) **TYPE OF DIVE:** Advanced–open water

A lonely and slightly forbidding-looking rock formation hides a treasure trove beneath its vertical cliffs: sharks. Blotting out the sunlight with their distinctive forms, a swarm of scalloped and great hammerheads patrol the waters overhead, often accompanied by hundreds of silky sharks.

Declared a UNESCO World Heritage site in 2006, Malpelo falls within the largest no-fishing zone in the eastern tropical Pacific Ocean. This is a sanctuary for large pelagics and a diver's dream, the opportunity to observe schools of large fish going about their natural behavior.

Hovering on the edge of very deep water (13,000 feet/3,962.4 m deep), Malpelo is a tiny, barren volcanic island with vertical cliffs, 300 miles (482.8 km) off the western Colombian coast, northeast of the Galápagos Islands. Malpelo's only inhabitants are large colonies of birds. Few divers know about Malpelo and even fewer make the trip, but the rewards for making the journey are evident as soon as you enter the water.

Strong and changing currents pull plankton closer to the surface. Although the occasional plankton

A yellow grouper swims among blacknosed butterflyfish.

What You'll See: Various Species of Sharks • Bigeye Trevallies • Red Snapper • Yellowfin Tuna • Pacific Creolefish • Eagle Rays • Manta Rays

deluge can reduce visibility from 100 feet (30 m) down to 30 feet (9 m), it makes for nutrient-rich waters that bring in bait balls, eagle rays, manta rays, tuna, whale sharks, and even a migrating whale or two. The rare smalltooth sand tiger shark and even rarer short-nosed ragged-tooth shark can be seen here, one of the few places in the world divers can spot these deep-water sharks. Butterflyfish also operate cleaning stations in the area, providing good viewing (and photographic) opportunities.

Malpelo isn't suitable for beginner divers. An advanced open water certification is the basic requirement, and divers need to have experience with currents and exceeding depths of 130 feet (39.6 m).

Politics also make Malpelo a difficult dive. Colombia doesn't have the best diving reputation and, at the time of writing this, the Colombian government recently changed the rules to ban foreign operators. Not a lot of operators are coordinating diving in the area, so do your research. Only one liveaboard is allowed in the marine park's borders at any given time, which provides an uncrowded dive experience, but you'll have to book well in advance.

Liveaboard vessels depart from Buenaventura, a Colombian town with a bad reputation, before steaming 30 to 40 hours to reach Malpelo. Diving is year-round, with water temperatures around 60 to 75°F (15.6 to 23.9°C). January to May is the best time for hammerheads, while June to December is calmer with better visibility.

Divers report that the hassles are worth it. This underwater ridge, with its steep walls, pinnacles, and caverns, is a spectacular, sharky sanctuary that only a few privileged divers get to experience. ○

"Diving with hammerheads is almost surreal; it's hard to believe what you're looking at. They're hypnotic, a little bit primal. You can imagine what the ocean was like 1,000 years ago when they're around."

—BRIAN SKERRY

Reaching polyps from an orange cup coral

Leatherback bass mate once a year in Malpelo.

Malepo is home to many bird colonies, including the masked booby.

The eerie, distinctive silhouettes of hammerheads draw divers to Malepo.

RUSSIA

WHITE SEA

Adventure diving under the sea—with beluga whales!

AVERAGE WATER TEMP: 28°F (−2.2°C) **AVERAGE VISIBILITY:** 18 to 100 feet (5.5 to 30 m)
AVERAGE DEPTH: Less than 70 feet (21 m) **TYPE OF DIVE:** Advanced—ice dive

Another world comes to life under the ice of the White Sea. The surging tide rises and falls, cracking, refreezing, and recracking the ice into spectacular, impermanent shapes. On this dive, you'll literally submerge yourself below the ice, after carefully cutting a human-size hole into the floes.

The frozen expanse you just broke into hides kelp forest nurseries, home to nudibranchs, soft corals, sea anemones, Gorgon's head brittle stars, sea stars, crabs, and northern wolffish. Suspended in the green-tinged water, jellyfish float like clouds.

Keep your eyes peeled, because the White Sea is one of the only places in the world you can dive with beluga whales all year round. If you're lucky enough to see one of these social creatures, you'll find them inquisitive, sometimes playful. Sometimes called sea canaries, belugas are one of the more vocal of whales.

Belugas and ice formations are the big draw here, but you'll see plenty more if you can tear your eyes away—and endure the cold a little bit longer. Diving beneath the ice is an alien experience. The floes cast a greenish tint to the surrounding water, and there's plenty of life to see.

Frilled anemones, nudibranchs, and a plethora of small invertebrates have found a way to endure the freezing temperatures and make a life here. If you find rock formations, peek

"During the melting season, sea ice breaks into pieces. These are not friendly little ice cubes, swimming through sea ice is like trying to swim against the current in a river full of loose concrete."

—ERIKA BERGMAN, SUBMARINE PILOT AND NATIONAL GEOGRAPHIC EXPLORER

Diving in the White Sea provides a rare chance to swim with belugas.

Anemones thrive in the cold White Sea.

What You'll See: Beluga Whales ● Ice Formations ● Seals ● Kelp Forests ● Temperate and Arctic Species

inside to uncover wolffish and bottom-dwelling Arctic sculpin and sea stars, as well as worms and spiny sun stars.

Look while you can—these rope-dependent dives don't last long (20 to 30 minutes max, if you can bear it) due to the torturous temperatures.

The White Sea is a peaceful and unique environment, one that few divers have explored, a true ice diving adventure.

This southern extension of the Barents Sea, tucked in between the left-hand corner of Russia and the right-hand corner of Finland, is a wild environment, with harsh conditions, virgin forests, and a wide expanse of tundra. It takes some time to get here, which is part of the adventure.

Not a lot of operators are coordinating diving in the area, so do your research. Ice diving and dry suit experience (with hood and dry gloves) are musts. Water temperatures hover around 28°F (−2.2°C) and visibility averages 18 feet (5.5 m), often dropping lower. Although it's a shallow dive, divers shouldn't be claustrophobic; you need to be confident you can remain calm underwater. ❍

Travel Tip:

Getting to Russia's Karelia region takes days, not hours, and is accessible via Finland, Moscow, or St. Petersburg. The best time for ice diving is late February to mid-April. Closer to February, it's darker and colder, but the ice is firm.

Rocky volcanic cliffs line Ascension's coast.

BRITISH OVERSEAS TERRITORY

ASCENSION ISLAND

Untouched, fascinating, subtropical paradise

AVERAGE WATER TEMP: 75°F (23.9 °C) **AVERAGE VISIBILITY:** 100 feet (30 m)
AVERAGE DEPTH: 25-plus feet (7.6+ m) **TYPE OF DIVE:** Advanced—rocky reef

Ascension Island is a young slip of land, the tip of a volcano that poked out of the water a million years ago. This dot in the South Atlantic, approximately halfway between South America and Africa, covers a scant 34 square miles (88 sq km) and has a fascinating history.

It first gained notoriety as a place of exile for passing sailors, a death sentence due to the lack of freshwater—the local cemetery is replete with victims of shipwrecks, disease, and maroonings. In 1815 the British garrisoned the island to prevent the rescue of perhaps the world's most famous exile, Napoleon, who had been banished to Ascension's closest neighbor, St. Helena, located 700 miles (1,126.5 km) southeast.

St. Helena residents described Ascension Island as a "cinder" to a visiting Charles Darwin, who

What You'll See: Dolphins • Yellowfin Tuna • Rockhind Groupers • Atlantic Blue Marlin • Black Triggerfish • Spotted Moray "Conger" Eels

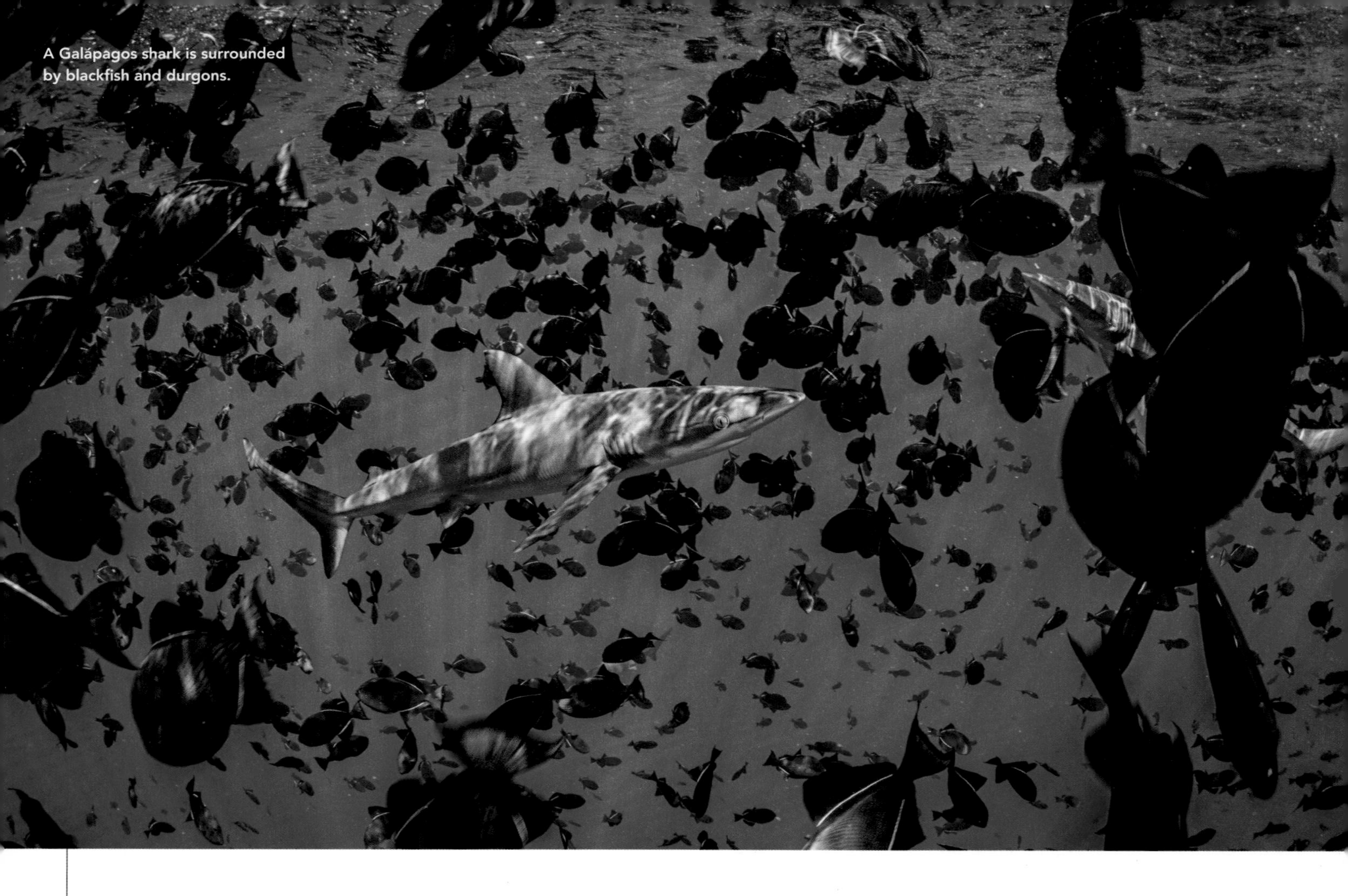

A Galápagos shark is surrounded by blackfish and durgons.

stopped here in 1836, resulting in one of the first terraforming experiments on record.

Ascension's point of pride used to be the World's Worst Golf Course (with oily, crushed-lava "greens"), but now it can rest its laurels on a marine reserve nearly as big as the United Kingdom.

Due to its remoteness, Ascension doesn't feature a wealth of marine diversity, but what it does have are fish in staggering abundance. The island hosts at least 11 endemic fish species (including the Ascension goby, resplendent angelfish, and Lubbock's yellowtail damselfish) and two species of shrimp (*Typhlatya rogersi* and *Procaris ascensionis*) that aren't found anywhere else in the world.

"The more we see, the more we realize what we don't know and for explorers that's the ultimate. Ninety-five percent of the deep ocean remains unexplored. We very much live on a water planet."

—BRIAN SKERRY

The more familiar fish that populate this area are surprisingly similar to the Caribbean: moray eels, groupers, trumpetfish, and soldierfish. The western coast attracts larger marine life like tuna, sharks, manta rays, dolphins, and the occasional whale. Green turtles are also frequent visitors to Ascension, laying their eggs between December and June.

Corals and sponges are few in a volcanic landscape that features plenty of caves, cavities, and a few wrecks.

A bottlenose calf swims by with its mother.

Travel Tip:

Travel to Ascension is changeable, so it's best to check with the Ascension Island government before getting too deep into planning. Visitors used to be able to catch a ride with the Royal Air Force, although the twice-weekly flights from Falkland Islands were suspended at the time of writing. However, there are weekly flights from South Africa via St. Helena. The R.M.S. *St. Helena* also visits Ascension approximately once every three weeks from Cape Town. Entry visas are required before arrival.

Divers lucky enough to explore Ascension all report feeling like they were the first ones to discover the site. The water is clear with up to 100 feet (30 m) of visibility, unpolluted, with sweeping cold currents (a 3mm wet suit is recommended for water temperatures ranging 70 to 80°F/21 to 26.7°C).

So why isn't Ascension overrun as a dive destination? It's remote, for starters, and it's not easy to get here. To dive Ascension, at the time of writing this, you have to join the local dive club (don't worry, their rates are reasonable) and bring all of your own gear. (Cylinder fills are available from the club.) With no official organization, most divers hire a local with a boat, but keep in mind that this is a working island, and most people are busy during the week.

It's also important to remember that Ascension isn't easy to dive, either. It has a wide variety of sites (shallow to deep, boat or shore dives), but swells can make shore diving risky, particularly in February and March. Divers report beginning their shore dives in flat calm waters, only to battle back to shore a half hour later. With help being 1,240 miles (1,995.6 km) away, it's essential to utilize local knowledge here. Even with swimming and snorkeling, a benign-looking beach can have hidden dangers.

The reason to go out of your way to visit Ascension is simple: You have the opportunity to be an explorer, an early adventurer in an area not overrun with tourists, a pristine marine environment that we hope will (thanks to its marine reserve protection) remain that way for many years to come. ❍

A trifecta of mahimahi

SOLOMON ISLANDS

DEVIL'S HIGHWAY

A mask-ripping wild ride with manta rays

AVERAGE WATER TEMP: 83°F (28.3°C) **AVERAGE VISIBILITY:** 50 feet (15 m)
AVERAGE DEPTH: 22 to 39 feet (6.7 to 11.9 m) **TYPE OF DIVE:** Advanced—drift

Kick down the wall, find a place to hang on (don't forget to wear gloves), and enjoy the wild ride. The Devil's Highway, also affectionately called the Devil's Washing Machine, is an advanced dive, not for the fainthearted, in strong eight- to 10-knot currents, but the payoff is a resident population of manta rays that block out the sun as they cruise overhead.

The raging currents act as a sushi train, serving up a steady stream of nutrients that keep the mantas (and fish) hovering around. With hard rock walls, the channel provides good visibility, and if you're good on air consumption, spend a little time exploring the cuttlefish-filled reef at the channel's exit.

This is a dive where you need to go with the flow, but also follow instructions, as the down currents can be fierce. And a little reef-hook experience wouldn't go astray before you chuck yourself in the washing machine.

A couple of liveaboards with years of experience navigate the Devil's Highway. Liveaboards depart from Honiara to the Florida Island group, which includes Mangalonga Island, where the famed highway can be found. Known as the Iron Bottom Sound, this stretch of water is littered with wrecks from a bloody, prolonged military campaign during WWII. Despite its wartime notoriety, the Solomon Islands are a laid-back, welcoming, untouched paradise.

With its equatorial climate, the Solomon Islands have a definite wet season from November to April, bringing up to nine feet (2.7 m) of rain, which is also the best time for encountering manta rays. June through August tends to be the coolest time of the year, although the average water temperature hovers around 83°F (28.3°C).

What You'll See: Manta Rays • Batfish • Cuttlefish • Scorpionfish

A hairy squat lobster hides
in its surroundings.

What You'll See: Top Wrecks • Gray Reef Sharks • Whitetip Reef Sharks • Tuna • Marlin

aircraft carrier up close, in detail, in a contained site. This wreck-filled lagoon is so unusual it became a UNESCO World Heritage site in 2010.

Bikini Atoll was opened to divers in 1996, but diving Bikini is still comparable to climbing Mt. Everest in the 1970s: It's expensive, remote, and dangerous. It's a 24-hour boat trip to help and, although visibility is crystal clear (130 to 200 feet/39.6 to 61 m), the depths require advanced skills—most wrecks rest around 150 feet (45.7 m), so deep-dive skills are a must. The majority of operators here recommend technical and wreck diving experience, as well as advanced Nitrox and deco training.

For those with the skills, the descent into the blue reveals ghost ships beyond imagining, like the 888-foot (270.7 m) U.S.S. *Saratoga,* a U.S. aircraft carrier lying upright in 180 feet (54.9 m) of water, or the H.I.J.M.S. *Nagato,* Admiral Isoroku Yamamoto's flagship for the attack on Pearl Harbor. Selected as a target ship for the nuclear blasts, the *Nagato* lies inverted, with her four propellers the most prominent features. Other wrecks include U.S. Navy submarines and the battleship U.S.S. *Arkansas.* Most are free from coral.

Several operators organize two-week dive trips to Bikini, but even the best-laid plans can be scuppered by airline reliability issues, increased oil prices, or other issues that affect the area. Once at Bikini, most liveaboards organize two deep dives a day, with an additional reef dive at a shallower depth if time and nitrogen levels permit.

The other reason Bikini is the holy grail for diving is its flourishing marine life, which has been untouched for 60 years. Sharks, especially, have rebounded, thanks in part to the creation of one of the world's largest shark reserves (768,547 square miles/1.99 million sq km) in 2011. ○

"There's a peace that comes from spending so much time in the ocean. There's an ebb and flow, a sense that you're part of something so much greater. There's a calming effect from being a diver and explorer and seeing the things we're privileged to see."

—BRIAN SKERRY

A brown booby dives for food.

A green turtle swims the waters of the Marshall Islands.

A diving helmet from the bridge of U.S.S. *Saratoga*.

A sunken bomber plane lies on the U.S.S. *Saratoga*'s port side.

BELIZE

SOUTH WATER CAYE

One of the world's best all-rounders

AVERAGE WATER TEMP: 82°F (27.8°C) **AVERAGE VISIBILITY:** 100-plus feet (30+ m)
AVERAGE DEPTH: 30 to 75 feet (9 to 22.9 m) **TYPE OF DIVE:** All level—open water

It's been called the best all-around diving destination in the world. South Water Caye is approximately 50 miles (80.5 km) southwest of the Great Blue Hole (page 384), its more famous countryman, but many divers consider this site to be the real crown jewel in the 185-mile-long (297.7 km) Belize Barrier Reef Reserve System.

Located in the largest protected marine area in Belize, South Water Caye is quieter, and rich with diversity. With visibility extending 100 feet and warm year-round water temperatures of 79 to 85°F (26 to 29.4°C), this spot has something for everyone.

Close to the island are sea grass beds extending to shallow lagoons that rarely dip deeper than 80 feet (24.4 m). This rich environment is home to corals (brain, sea fan, black), sponges, and reef fish, making it an easy exploration for beginner divers.

Intermediate divers also have plenty of entertainment on offer. Close to the caye and extending eight miles south, an underwater cliff face starting around 40 feet (12 m) drops off into the deep blue, providing the perfect habitat for sharks, tuna, barracuda, groupers (including the endangered Nassau and black), whale sharks, and rays.

Located 16 miles (25.7 km) off the coast from Dangriga, this tiny 15-acre (0.06 sq km) island has two resorts, a dorm for visiting students, and nothing else. (It can also be accessed via day trips or by liveaboard.)

The best time to visit is during the dry season, November to April. Located approximately 75 miles (120.7 km) south of Belize City, Dangriga can be reached via bus, car, or a short flight, and from there it's a 40-minute boat ride to South Water Caye. ❍

What You'll See: Brain, Staghorn, and Finger Coral • Sea Whips • Sea Fans • Black Coral • Sponges • Reef Fish • Tawny Nurse Sharks • Tuna • Barracuda • Nassau and Black Groupers • Whale Sharks • Spotted Eagle Rays • Southern Stingrays • Hawksbill Turtles • Spiny Lobsters • Spider Crabs

A green moray eel
surrounded by mysids

Tabular icebergs as large as buildings float in Antarctica's waters.

ANTARCTICA

The most remote and extreme diving experience in the world

AVERAGE WATER TEMP: 30°F (−1.1°C) **AVERAGE VISIBILITY:** 30 to 60 feet (9 to 18 m)
AVERAGE DEPTH: Less than 82 feet (25 m) **TYPE OF DIVE:** Advanced—cold water and ice

"The waters under Antarctic ice are like Mount Everest: magical, but so hostile that you have to be sure of your desire before you go. You cannot go halfheartedly; you cannot feign your passion. The demands are too great. But that's what makes the images you see here unprecedented, and the experience of having taken them and of having seen this place so unforgettable." So wrote French biologist and photographer Laurent Ballesta in a July 2017 story for *National Geographic* magazine.

Although Antarctica tourism is on the rise, few people brave scuba diving in this remote, extreme environment. The cold is painful—so painful it dictates dive time (around 20 or 30 minutes max), rather than air supply. Familiarity with and use of a dry suit, hood, and gloves are essential, though they won't keep the cold at bay.

This should be expected from the landscape that recorded the lowest temperature on the

What You'll See: Leopard Seals • Penguins • Soft Corals • Sea Stars • Octopuses • Ice Formations • Kelp Walls • Spider Crabs • Sea Snails

Hunting orcas crest the Antarctic waters.

planet—minus 128.56°F (−89.2°C) in July 1983 at the Russian research outpost of Vostok. Sea temperatures range from 28 to 34°F (−2.2 to 1.1°C), but the high salinity keeps the water from freezing. Antarctica is the world's biggest desert, at 5.4 million square miles (14 million sq km), 98 percent of which is covered with ice, receiving a scant two inches (5 cm) of annual rainfall. It is high, dry, and windy.

The joke is that the only typical Antarctic dive is a cold one. What the barren landscape conceals is a lush garden of life, utterly unique due to millions of years of separation. Curious and impressive leopard seals and agile penguins are regular dive companions. Marine life such as kelp walls, jellyfish, sea stars, spider crabs, soft corals, anemones, and sea snails also flourish under the ice.

The way some creatures have adapted to this hostile environment is fascinating: Antarctica is home to at least 16 species of octopuses, which survive subfreezing temperatures thanks to a specialized pigment in their blood. Bottom-dwelling icefish (50-plus species) also have proteins in their blood that help them survive in the cold. There is even an anemone species known to live *in* the ice.

"Like others, I went to Antarctica for the animals but I return for the ice. The Austral Ocean is a polar garden filled with iceberg sculptures that are a perfect metaphor for the sea."

—DAVID DOUBILET, UNDERWATER PHOTOGRAPHER

While diving along the ice floes, the sculpture and endless colors, refracted and lit by sunlight, are dizzyingly beautiful, an unforgettable experience in an equally unforgettable land.

A few cruise vessels offer diving in Antarctica, varying from shallow ice diving to shore diving, with consistent visibility of 30 to 60 feet (9 to 18 m). Conservative limits are observed due to the distance from emergency aid. ❍

Divers explore the fascinating underwater ice formations.

A Weddell seal pup nudges its mother beneath the ice.

TRINIDAD AND TOBAGO

SISTERS ROCKS

A dramatic drift dive with hammerheads

AVERAGE WATER TEMP: 78°F (25.6°C) **AVERAGE VISIBILITY:** 60 to 80 feet (18 to 24 m)
AVERAGE DEPTH: 40 to 75 feet (12 to 22.9 m) **TYPE OF DIVE:** Advanced—drift

A cluster of five rock pinnacles rise from a depth of 130 feet (39.6 m) in the Southern Caribbean, surrounded by ocean-floor rubble of boulders and rocks, cut through with sand chutes. Here, a regular population of majestic hammerheads prefers to cruise, easily cutting through strong currents and swells the site is famous for.

Known as the Sisters Rocks, this dive site is located two miles (3.2 km) offshore of Tobago's Bloody Bay, a name that highlights this dual-island nation's colorful history. For years the Dutch, English, and French battled over Tobago. The island changed hands at least 30 times before the British finally laid claim in 1814.

Due to its location (off the northern coast of South America, 10 miles/16 km north of Venezuela), the Sisters Rocks are a consistent place to spot hammerheads (great and scalloped) and other large pelagics, such as sharks (reef, sand tiger, hammerheads), tarpon, and barracuda.

The challenge of this dive site (the strong current) is also the reason it's known as the Drift Dive Capital of the Caribbean. Unpredictable weather and swells can quickly render the conditions undiveable, but on a good day, with a swift ride and schooling hammerheads, it's easy to understand this dive's popularity.

December through March, which is conveniently the dry season, is also the best time for spotting the resident hammerheads. Temperatures are pleasant throughout the year, with a median water temperature of 78°F (25.6°C) and consistent 60 to 80 feet (18 to 24 m) of visibility. ❍

What You'll See: Great Hammerheads • Scalloped Hammerheads • Tarpon • Barracuda • Rock Formations

A giant crinoid feather star stretches its arms.

BELIZE

GREAT BLUE HOLE

A journey to the center of the Earth

AVERAGE WATER TEMP: 82°F (27.8°C) **AVERAGE VISIBILITY:** 100-plus feet (30+ m)
AVERAGE DEPTH: 130 feet (39.6 m) **TYPE OF DIVE:** All level—sinkhole

You'll want to start this dive with a flight. The best way to absorb the breathtaking beauty and gargantuan size of the Great Blue Hole is to get a bird's-eye view. An aerial tour will fly you over Lighthouse Reef so you can marvel at the circular, blue abyss that seems to lead directly to the heart of the ocean, whetting your appetite for the dive to come.

This sapphire blue, underwater sinkhole is more than 984 feet (300 m) across, with a depth of 410 feet (125 m). The dive is surprisingly simple—just deep. Follow your guide, descending quickly to 130 feet (39.6 m). Here, the cavern ceiling molds into a row of stalactites, some of which are three feet (0.9 m) wide. You'll have to marvel quickly, because you'll have only eight to 12 minutes at that depth before beginning your ascent.

This 70-million-year-old flooded cave is part of the Belize Barrier Reef Reserve System, the largest barrier reef in the Northern Hemisphere and a UNESCO World Heritage site. (In 2018, the coral reef world celebrated a hard-won victory when UNESCO removed the Belize Barrier Reef from its list of endangered World Heritage sites.) It's fantastic news that Jacques Cousteau, who frequently listed the Great Blue Hole as one of the world's top 10 dive sites, undoubtedly would have celebrated. The Great Blue Hole's magnificent visibility—up to 100 feet (30 m)—is still its key feature, but there is also plenty to see, from the sinkhole formations and stalactites to the circling reef and bull sharks.

The Great Blue Hole is approximately 60 miles (96.6 km) off the coast from Belize City. This popular dive site is easily accessed by day trips from the mainland or from liveaboards. The best time to visit is in the dry season, November to April. Year-round water temperatures range from 79 to 85°F (26 to 29.4°C). ❍

What You'll See: Sinkhole Formations • Stalactites • Reef Sharks • Bull Sharks

Before you dive, take in the
Great Blue Hole from the sky.

DIVER CONSERVATION

Apathy. It's the biggest issue facing marine conservation. Feelings of helplessness, of being overwhelmed, of not wanting to know—they are as threatening as oil spills. But we cannot blind ourselves. The ocean is in peril. Think of what we've done to it already: a human history of trash; pollutants; and plastic, radioactive, and chemical waste heaped into the sea. The ocean has served as our convenient, secret garbage dump. Because we don't see the damage, we are able to turn a blind eye. But no longer. Plastic alone is estimated to kill millions of marine animals every year—more than 700 species, including endangered ones, are affected by it. And that's just plastic. There's also a massive decline in our coral reefs and fish populations from other harmful combatants—from sunscreen to overfishing to climate change.

We still have time to act, but the time to make the *decision* to act has passed.

"Divers have not always been great conservationists," says Brian Skerry, "but there's a wonderful, positive evolution among divers who get it, and that's progress. Divers evolve. You can't have dived for any length of time without becoming a little bit wiser about your environment. We've lost so much: We've essentially experienced geologic change in our lifetime."

Divers should be leading the charge to protect the ocean because we are in a privileged position to actually see change taking place. It's our backyard, our playground, our responsibility.

Not sure where to begin? Here are seven guidelines to get you started:

1. **Vote.** This is an easy—and important—one. Vote for politicians who champion conservation and let them (and their opponents) know why. Be vocal. For as much good as we're doing individually, we need lawmakers to back it up on a global level.
2. **Support science.** Misinformation and choked-off funding is making scientific research difficult, right when we need it most. Whether it's a local grassroots project in your community, organizations looking at the larger picture, or

solution-focused companies, you can provide support through donating your time, volunteering your individual expertise (be it graphic design or social media know-how), or backing them financially. Get involved.

3. **Watch what you eat.** "We've wiped out 99 percent of codfish stocks since colonial England," Skerry says. "I see people—divers—ordering all kinds of seafood and they're not careful about what they're ordering." Familiarize yourself with overfished and threatened species and make a point to avoid (or limit) consumption.
4. **Use sunscreen smarts.** Most sunscreens contain chemicals like oxybenzone and octinoxate that are toxic to coral. Every year, about 14,000 tons of sunscreen find its way into the sea. (Even if you aren't swimming, the toxins still go down the drain.) Do your research and choose more coral-friendly, mineral-based sunblocks, or cover yourself with clothing, another effective sun-prevention method. Remember, your choice makes a difference—it all adds up.
5. **Let your tourism dollars do the talking.** Your dollars—and where you spend them—matter. Take sharks, for example. A living shark can generate more than five million dollars in ecotourism throughout the course of its life, while its fin (and death) nets only $200. An estimated 100 million sharks are killed every year—we're strafing the ocean of these essential predators, and the only heavyweight industry that has a hope of tackling the shark-finning and bycatch industries is tourism. "If you pick conservation operators, there will only be conservation operators," says Dr. Andrea Marshall, a manta ray scientist and conservationist. "Your choices influence the industry, which is why it's important to support places that are putting protections in place."
6. **Look but don't touch.** This should be a no-brainer, but in the age of Instagram, we could all use another reminder: Let wildlife choose the interaction. Give animals at least six feet (1.8 m) of space, and do not touch marine life. Avoid (and report) any operation that doesn't apply good practice or alters the behavior of wildlife (removing the tails of stingrays, for example, so they don't pose a danger to swimmers). In this book, we have recommended a few destinations—shark diving in Port Lincoln and Guadalupe, for example, or night diving with manta rays in Kona—that do bait or otherwise influence interactions, but these have been carefully considered and included for two reasons: First, the area has quality

operators, a good history, and established rules and regulations. Second, the benefit of these interactions outweighs the slight behavioral change. If divers see a great white shark up close, they are more likely to care about sharks. That caring leads to conservation. Places like Beqa Lagoon and Tiger Beach have cultivated a legion of shark ambassadors. That being said, do your homework, and go in with your eyes open. The safety of divers and animals should always be paramount to the experience.

7. **Remember that plastic is not so fantastic.** In May 2018 National Geographic launched its Planet or Plastic? campaign, a multiyear effort to raise awareness about the global plastic crisis. A few fast plastic facts for you:
   - More than 40 percent of plastic is used once and then discarded, and 6.3 billion tons (5.7 billion metric tons) of plastic fill our landfills, landscapes, and oceans.
   - Nearly 700 marine species are known to be affected by ocean plastic, which kills millions of ocean animals every year.
   - Nearly one million plastic bottles are sold every minute around the world.

   Be mindful of the amount of plastic you use in your everyday life and take active steps to reduce it. You can do this while diving by bringing reusable water bottles with you when you travel, as well as multiuse utensils that you can pull out instead of the plastic stuff when you need a bite to eat on your dive boat or airplane. Oh, and skip the plastic straws, please. It's that simple. And that effective.

Every choice you make is harmful or helpful. Even small steps—especially small steps—make a difference. *You* are the solution to marine conservation.

> *"Once you become aware of the issues, solutions will present themselves. Everything is connected. Everything really matters and we can no longer look at things in a vacuum."*
>
> —BRIAN SKERRY

DIVE DESTINATIONS BY COUNTRY

DIVER RESOURCES

With millions of active divers worldwide, the scuba diving community is a dynamic, engaged, and global network, with a wide variety of resources available to those interested in certification, continuing education, conservation, travel, or connection with like-minded diving enthusiasts.

Professional Association of Diving Instructors (PADI)

PADI is the world's leading scuba diver training organization, with nearly 6,500 PADI Dive Centers and Resorts around the world and more than 135,000 PADI professional members. Since its establishment in 1966, PADI has issued more than 25 million scuba diving certifications. PADI courses offer diver certification and education programs that introduce skills, safety-related information, and local environmental knowledge to student divers at every stage. Committed to being a force for good, with every diver certification issued PADI also creates a growing number of ocean ambassadors to help defend and preserve the marine environment. *padi.com*

Mission Blue

Led by legendary oceanographer Sylvia Earle, Mission Blue is bringing together a global coalition to inspire public awareness, access, and support for marine protected areas named Hope Spots. These special places are vital to the health of our ocean. There are currently 95 designated Hope Spots, and nominations for new areas are always welcome. Currently, the Mission Blue alliance includes more than 200 ocean conservation and like-minded organizations. Mission Blue also offers exploration and education expeditions to these key oceanic regions. *mission-blue.org*

Pristine Seas

In 2008, National Geographic Explorer-in-Residence and marine ecologist Enric Sala launched a project to explore and protect the ocean's wild places. Expeditions involve weeks at sea and thousands of diving hours, in an effort to explore, understand, and protect areas where marine life can thrive, encouraging stewardship and management of these areas. The result? The establishment of some of the largest marine reserves in the world. *nationalgeographic.org/projects/pristine-seas*

Conservation Organizations

Environmental stewardship—particularly of the ocean—has never been more important. Approximately 12 percent of land around the world is under a form of protection, but less than 6 percent of the ocean is protected. Here are some organizations that are rolling up their sleeves and getting involved: SeaLegacy (*sealegacy.org*): a collective of some of the world's best storytellers, filmmakers, and photographers working at #TurningtheTide for healthy and abundant oceans. Sea Shepherd (*seashepherd.org*): Established in 1977, it has a proactive mission of protecting ecosystems and species by documentation, investigation, and action. 4Ocean (*4ocean.com*): organizes cleanup events and employs teams of cleanup crews around the world, seven days a week. The distinctive 4Ocean glass-bead bracelets, made from recycled materials, represent one pound of removed ocean trash. Take 3 (*take3.org*): an Australian nonprofit that promotes a simple message: Take three pieces of trash with you whenever you leave a beach or waterway—an easy way to make a difference. Lonely Whale (*lonelywhale.org*): a facilitator for innovation, ideas, and collaboration to drive impactful change for the oceans. One of its most powerful campaigns is the Strawless Ocean initiative encouraging us to #StopSucking. Planet or Plastic? (*nationalgeographic.com/environment/planetorplastic*): National Geographic's multiyear effort to raise awareness about the global trash crisis, including education and information about how you can reduce your reliance on single-use plastics.

Divers Alert Network (DAN)

DAN is synonymous with scuba diving safety. Founded in 1980, DAN helps divers in need of medical assistance and promotes dive safety through education (everything from case studies to online resources to courses), research, and diving services including dive accident insurance plans. It is supported by membership dues and donations. Alert Diver (*alertdiver.com*), DAN's magazine, is filled with articles ranging from safety and dive medicine to travel and conservation. It's available to DAN members and is also published for free on DAN's website, which has a wealth of information. *diversalertnetwork.org*

National Oceanic and Atmospheric Administration (NOAA)

Facts can be difficult to come by these days, but if you're looking for general information about the changing environment, NOAA is a good place to start. From climate monitoring to coastal restoration to weather forecasts, NOAA's purpose is to understand and predict changes in climate and oceans, share that knowledge and information, and conserve and manage resources. *noaa.gov*

Online Networks

Digital resources for divers abound, with a little something for everyone. Underwater Photography Guide (*uwphotographyguide.com*) is for divers passionate about capturing the best shot, featuring gear reviews, tutorials, and information for photographers of any level, from beginner through professional. Girls that Scuba (*girlsthatscuba.com*), the largest community for female divers, is exploding in popularity, providing travel information, dive articles, and education and support for female divers around the globe. Undercurrent (*undercurrent.org*) is a nonprofit that has been publishing a monthly, subscription-only newsletter since 1975. It's known to be forthright, honest, and ad-free.

Trade Shows

Diving trade shows are a great opportunity to travel, meet fellow divers, and learn more about the diving industry and opportunities. DEMA (*demashow.com*) is one of the largest and most well known, attracting thousands of dive and travel industry professionals from around the world. The Boston Sea Rovers (*bostonsearovers.com*) is one of the granddaddies. Founded when scuba diving was in its infancy, it offers a diverse and fascinating annual show. BOOT (*boat-duesseldorf.com*) is an international watersports trade fair that has been running since 1970. ADEX (*adex.asia*), the largest and longest running dive expo in Asia, dedicates each show to a specific theme (sharks or climate change, for example).

Scuba Schools International (SSI)

Started in 1970, SSI offers scuba diving training, certification, and education programs for beginners through professionals. The global network of divers, dive instructors, and dive centers extends over 2,800 international locations in more than 110 countries. *divessi.com*

British Sub-Aqua Club (BSAC)

Founded in 1953, BSAC is the United Kingdom's national governing body for scuba diving, offering internationally recognized diver training and social activities, with a wide network of clubs and diving centers. The organization aims to promote the sport of diving and protect the U.K.'s underwater and wreck heritage through education, skills training, and collaborative programs with groups like the Marine Conservation Society. *bsac.com*

American Canadian Underwater Certifications, Inc. (ACUC International)

Committed to safety, education, and conservation, ACUC is a certification organization with programs in more than 43 countries around the world. ACUC offers all levels of certification for recreational scuba diving, from beginner to instructor. *acuc.es/en*

World Underwater Federation (CMAS)

Founded in 1959, the World Underwater Federation includes 130 federations from five continents. Along with being at the forefront of scientific and technical research and development, it also organizes international underwater sporting events and has established an extensive diver training and certification system. *cmas.org*

National Association of Underwater Instructors (NAUI)

NAUI has a reputation for technical training and conducts research on the cutting edge in this field. Its diver education programs have been used by U.S. Navy SEAL teams and NASA's Neutral Buoyancy Laboratory, where astronauts train for their space walks. NAUI is also committed to preservation and conservation through its Green Diver Initiative, empowering individuals to conserve and preserve the ocean environment. *naui.org*

DEDICATION

For Oliver Payne, who values tenacity. Thank you for opening the door to this mad, wonderful National Geographic adventure and altering the course of my life.

And Chris Taylor, who should have been born with gills.

ACKNOWLEDGMENTS

Special thanks to Chris for your patience and support while I found my way to diving. (Just try getting me out of the water now!) You're my favorite person to explore with, both above and under the water.

And also to *National Geographic* magazine editor Oliver Payne for taking the time to respond to a college kid consumed with writing for National Geographic. If you hadn't answered my letter, my life would have been very different. You are one of my navigation points, a fellow connoisseur of good stories well told, and a friend. Thank you for 20-plus years of quiet, consistent support.

My family and friends have watched me drive myself mad with my writing and—for some odd reason—love and support me all the same, which makes all of you crazier than I am. And I love you for it.

Allyson Johnson, senior editorial project editor extraordinaire, what a ride! Thank you for your guidance, encouragement, patience, and friendship. (Do you realize we created this book throughout 16 countries?) Love working with you and looking forward to future projects.

Thank you to the entire National Geographic Books team for your help and hard work: Hilary Black, Bill O'Donnell, photo editors Moira Haney, Jill Foley, and Krista Rossow (my friend and partner in crime on several assignments), art director Sanaa Akkach, senior production editor Judith Klein, copy editor Heather McElwain, financial wizards Pinar Taskin and Jeannette Swain, and countless others who helped on this endeavor. It really does take a village to bring a book to life, and the behind-the-scenes team never gets the recognition deserved for their dedication. Thank you.

I consulted a legion of divers for this project and I was overwhelmed by their passion for the ocean. Thank you for your help and advice. Best lists, by definition, are contentious, but I hope you like the end result and spot some of your favorites.

My thanks to the Nat Geo family of photographers and explorers, including Erika Bergman, Camrin Braun, Jess Cramp, David Doubilet, Sylvia Earle, David Gruber, Andy Mann, Andrea Marshall, Cristina Mittermeier, Erina Molina, Paul Nicklen, Jennifer Adler Owen, Andrea Reid, Enric Sala, and Shannon Swanson. I have admired your work for a long time and look forward to seeing what explorations and discoveries you embark on next.

A special thanks to Brian Skerry for spending hours sharing your stories, knowledge, and love of underwater exploration with me.

Last, to my extended Nat Geo family—especially you, George Stone—thanks for the inspiration, the support, and the wild ride.

ABOUT THE AUTHOR

Carrie Miller has been writing for National Geographic since 1998. This two-time Lowell Thomas Award winner and contributing editor at *National Geographic Traveler* magazine calls New Zealand home, but in 2018 she put everything in storage to travel the world with her diver husband, Chris Taylor, on a yearlong assignment exploring the world's 50 best dive travel locations. (The resulting book will be published by National Geographic in early 2020.) We actually have no idea where she is now, but near the ocean is a safe bet.

Follow along on Facebook (The Dive Travelers) and Instagram (@thedivetravelers). And find Carrie on Facebook (Carrie Miller Writer) and Instagram (@carriemiller_writer).

ILLUSTRATIONS CREDITS

Front cover, Alex Mustard/naturepl.com; 2-3, Gabriel Barathieu; 6-7, Laurent Ballesta/National Geographic Creative; 9, Brian Skerry; 11, Brian Skerry/National Geographic Creative; 13, Seanna Cronin; 14-15, Jason Lafferty; 18-19, Jurgen Freund/naturepl .com; 21, Karsten Wrobel/Getty Images; 23, Alex Mustard/ naturepl.com; 25, Luis Javier Sandoval Alvarado/Getty Images; 26-27, Shawn Heinrichs; 29, RGB Ventures/SuperStock/Alamy Stock Photo; 31, Pete Atkinson/Getty Images; 32-33, Gary Bell/ Oceanwide Images; 35 (LE), Gary Bell/Oceanwide Images; 35 (UP), Bill Boyle /Oceanwide Images; 35 (LO), Rudie Kuiter/Oceanwide Images; 36-37, James Peake/Alamy Stock Photo; 39, Terry Moore/Stocktrek Images/Getty Images; 41, Adriana Basques; 42-43, Franco Banfi; 45, nejdetduzen/Getty Images; 47, Tim Rock Photography; 49, Ethan Daniels; 50-51, Mirko Zanni/Getty Images; 53, Heather Perry/National Geographic Creative; 55, WaterFrame/Alamy Stock Photo; 57, Sam Cahir; 58-59, by wildestanimal/Getty Images; 60, by wildestanimal/Getty Images; 61, Brian Skerry; 63, Nadia Aly; 65, Reinhard Dirscherl/ullstein bild via Getty Images; 67, JTB Photo/UIG/age fotostock; 68-69, Hoiseung Jung/EyeEm/Getty Images; 71, Gabriel Barathieu; 73, Westend61/Getty Images; 74-75, Auscape/UIG/Getty Images; 77 (LE), Turnervisual/iStock/Getty Images; 77 (UP), Southern Lightscapes-Australia/Getty Images; 77 (LO), Mathieu Meur/Stocktrek Images/Getty Images; 78-79, Seanna Cronin; 81, João Paulo Krajewski; 83, Anthony Grote/Blue Safari Seychelles; 85, Alexander Safonov/Getty Images; 86-87, Sam Cahir; 89, Colin Marshall/ FLPA/Minden Pictures; 91, Alex Mustard; 92-93, Tobias Friedrich/ WaterFrame/age fotostock; 95, Jon Kreider Underwater Collection/Alamy Stock Photo; 97, Steve De Neef/National Geographic Creative; 98-99, Franco Banfi; 101, Matty Smith; 102-103, David Doubilet/National Geographic Creative; 104, David Doubilet/ National Geographic Creative; 105, Shawn Heinrichs; 107, Reinhard Dirscherl/ullstein bild via Getty Images; 109, WideScenes Photography/Helen Woodford; 101-111, Brian Skerry/National Geographic Creative; 113, David Doubilet/National Geographic Creative; 114-115, Scubazoo/Alamy Stock Photo; 116, Clark Miller; 117, Adriana Basques; 118, Daniel Selmeczi/stevebloom .com; 119, Jurgen Freund/naturepl.com; 121, Mauricio Handler/ National Geographic Creative; 122-123, Mauricio Handler/ National Geographic Creative; 125, Jenny & Tony Enderby/Getty Images; 127, 22August/Shutterstock; 129, Nadia Aly; 130-131, Reinhard Dirscherl/ullstein bild via Getty Images; 132, Nadia Aly; 133, Nadia Aly; 135, Nadia Aly; 137, Mike Veitch/Alamy Stock Photo; 139, Media Drum World/Alamy Stock Photo; 140-141, Jason Lafferty; 143(LE), Jason Lafferty; 143 (UP), David Fleetham/ VW PICS/UIG via Getty Images; 143 (LO), Franco Banfi/Getty Images; 144-145, Jason Lafferty; 147, Ethan Daniels; 148-149, Shawn Heinrichs; 151, Danielle Villasana/Redux; 153, Stocktrek Images/National Geographic Creative; 155, viper345/Shutterstock; 156-157, Joe Dovala/Getty Images; 159 (LE), Michael Stubblefield/Alamy Stock Photo; 159 (UP), Norbert Wu/Minden Pictures; 159 (LO), Elyn Stubblefield/Alamy Stock Photo; 160-161, Brandi Mueller/Getty Images; 163, Antonino Bartuccio/SIME/ eStock Photo; 165, Shawn Heinrichs; 166-167, Nadia Aly; 169, Gabriel Barathieu; 171, Paul & Paveena Mckenzie/Getty Images; 172-173, Global_Pics/Getty Images; 175, Corinne Bourbeillon; 176-177, Edwar Herreno; 179, lindsay_imagery/Getty Images; 180-181, Steve Daggar Photography/Getty Images; 182-183, Matty Smith; 185 (LE), mauritius images GmbH/Alamy Stock Photo; 185 (UP), Reinhard Dirscherl/Alamy Stock Photo; 185 (LO), mauritius images GmbH/Alamy Stock Photo; 186-187, Michael McCoy/Getty Images; 189, imageBROKER/Alamy Stock Photo; 190-191, Danita Delimont/Getty Images; 192, Tim Laman/ National Geographic Creative; 193, Jason Edwards/National Geographic Creative; 194-195, Jason Edwards/National Geographic Creative; 197, David Noton/naturepl.com; 199, Alex

Mustard; 200-201, Enric Sala/National Geographic Creative; 202, Enric Sala; 203, Brandi Mueller/Getty Images; 205, Daniela Dirscherl/Getty Images; 206-207, Philippe Bourseiller/Getty Images; 208, Jurgen Freund/NPL/Minden Pictures; 209, Reinhard Dirscherl/Getty Images; 211, João Paulo Krajewski; 212-213, Randy Olson/National Geographic Creative; 215, Seanna Cronin; 216-217, Seanna Cronin; 219, Secret Sea Visions/Getty Images; 220-221, Brian Skerry; 223 (LE), Paul Kay/Getty Images; 223 (UP), Brian Skerry; 223 (LO), Brian Skerry; 224-225, Paul Kay/Getty Images; 227, Franco Banfi; 228-229, Franco Banfi; 230, Alex Mustard; 231, Alex Mustard; 233, David Doubilet/National Geographic Creative; 234-235, Darren Jew/Australian Geographic; 237, Michael Runkel/Robert Harding; 238-239, Mark Strickland; 240-241, Joe Dovala/Getty Images; 243 (LE), Brandi Mueller/Getty Images; 243 (UP), Joe Dovala/Getty Images; 243 (LO), Brandi Mueller/Getty Images; 244-245, Brandi Mueller/Getty Images; 247, Stocktrek Images/National Geographic Creative; 248-249, Edwar Herreno; 251 (LE), Harald Slauschek/ASAblanca via Getty Images; 251 (UP), Matty Smith; 251 (LO), Matty Smith; 252-253, Ethan Daniels; 255, Overflightstock Ltd/Alamy Stock Photo; 256-257, Stuart McCall/Getty Images; 259, Tim Laman/National Geographic Creative; 261, Media Drum World/Alamy Stock Photo; 262-263, David Doubilet/National Geographic Creative; 265, Reinhard Dirscherl/ullstein bild via Getty Images; 266-267, Reinhard Dirscherl/Getty Images; 269, David Arbogast; 270-271, by wildestanimal/Getty Images; 272, Copyright Michael Gerber/Getty Images; 273, Bernard Radvaner/Getty Images; 274, Mauricio Handler/Getty Images; 275, Pete Oxford/Minden Pictures; 277, Franco Banfi; 278-279, Westend61/Getty Images; 280, Franco Banfi; 281, Franco Banfi; 282-283, Mikhail Starodubov/Shutterstock; 284, Joakim Boneng; 285, Alessandro Saffo/SIME/eStock Photo; 287, Brian Skerry; 288-289, Brian Skerry; 290, Jim Abernethy/National Geographic Creative; 291, Brian Skerry/National Geographic Creative; 293, Seaproof.tv; 295, Gabriel Barathieu; 296-297, Dan Burton/Getty Images; 299, Fred Bavendam/Minden Pictures; 300-301, Alexander Safonov/Getty Images; 303 (LE), Mauricio Handler/National Geographic Creative; 303 (UP), Mauricio Handler/National Geographic Creative; 303 (LO), Peter Pinnock/Getty Images; 304-305, Alexander Safonov/Getty Images; 306-307, Grant Henderson/Alamy Stock Photo; 309 (LE), Rodger Jackman/Getty Images; 309 (UP), Steven Maltby/Shutterstock; 309 (LO), Steven Maltby/Shutterstock; 310-311, Geraint Tellem/Alamy Stock Photo; 313, Jeremy Rodriguez; 315, Daniel Lamborn/Alamy Stock Photo; 316-317, Edwar Herreno; 319, Franco Banfi/Barcroft Media/Barcroft Media via Getty Images; 320-321, Reinhard Dirscherl/ullstein bild via Getty Images; 322, Borut Furlan/Getty Images; 323, Edwar Herreno; 324, Borut Furlan/Getty Images; 325, Edwar Herreno; 327, Jill Heinerth; 328-329, Laurent Ballesta/National Geographic Creative; 331, Fred Olivier/naturepl.com; 332-333, Jason Edwards/National Geographic Creative; 335, JP Bresser; 336-337, Mark Strickland; 339 (LE), Reinhard Dirscherl/ullstein bild via Getty Images; 339 (UP), Mark Strickland; 339 (LO), Mark Strickland; 340-341, iStock/Getty Images; 343, Franco Banfi; 344-345, Westend61/Getty Images; 346, Paul Nicklen/National Geographic Creative; 347, Christian Vizl/TandemStock.com; 349, Paul Nicklen/National Geographic Creative; 350-351, Edwar Herreno; 353 (LE), Franco Banfi/Getty Images; 353 (UP), Bernard Radvaner/Getty Images; 353 (LO), Borut Furlan/Getty Images; 354-355, Bernard Radvaner/Getty Images; 357, Franco Banfi; 358-359, Photo by Viktor Lyagushkin/Getty Images; 360-361, Prisma by Dukas Presseagentur GmbH/Alamy Stock Photo; 362, Ellen Cuylaerts; 363, Ellen Cuylaerts; 364, Kent Kobersteen/National Geographic Creative; 365, Ellen Cuylaerts; 367, Reinhard Dirscherl/ullstein bild via Getty Images; 368-369, Reinhard Dirscherl/FLPA/Minden Pictures; 371 (LE), Reinhard Dirscherl/Alamy Stock Photo; 371 (UP), Reinhard Dirscherl/Getty Images; 371 (LO), Reinhard Dirscherl/ullstein bild via Getty Images; 372-373, Reinhard Dirscherl/ullstein bild via Getty Images; 375, Tony Rath Photography; 376-377, Caitlyn Webster; 378, Krista Rossow; 379, Franco Banfi; 380-381, Laurent Ballesta/National Geographic Creative; 383, Doug Perrine/naturepl.com; 385, Brian Skerry/National Geographic Creative; Back cover, Jad Davenport/National Geographic Creative.